ALIAS J. J.
CONNINGTON

ALIAS J. J. CONNINGTON

A. W. Stewart

COACHWHIP PUBLICATIONS
Greenville, Ohio

Alias J. J. Connington
Copyright © The Professor A W Stewart Deceased Trust.
Alias J. J. Connington published 1947.
"After Death the Doctor" published in *The First Class Omnibus* (1934)
2015 Coachwhip edition
Introduction © 2015 Curtis Evans

ISBN 1-61646-322-8
ISBN-13 978-1-61646-322-9

Cover elements: Symbols © Shoo Arts; Gun © Blan-K.

CoachwhipBooks.com

All Rights Reserved. No part of this publication may be reproduced, stored in a retrieval system or transmitted in any form or by any means—electronic, mechanical, photocopy, recording or any other—except for brief quotations in printed reviews, without the prior permission of the author or publisher.

CONTENTS

Introduction, by Curtis Evans	7
Preface	23
1 · The Mystery of Chantelle	29
2 · Where Plots Come From	54
3 · Steps Towards the Atomic Bomb . . .	70
4 · How To Poison Your Intellect	99
5 · Money for Nothing	120
6 · The Wonderful Lamp	149
7 · What is Your Evidence Worth?	181
8 · Elephants for Want of Towns	209
9 · The Manichee	222
10 · A Present for a Good Historian	253
11 · Novelties in Their Day	271
12 · Things in the North	293
"After Death the Doctor"	327

INTRODUCTION
CURTIS EVANS

Alfred Walter Stewart (1880-1947)
Alias J. J. Connington

DURING THE GOLDEN AGE of the detective novel, in the 1920s and 1930s, "J. J. Connington" stood with fellow crime writers R. Austin Freeman, Cecil John Charles Street, and Freeman Wills Crofts as the foremost practitioner in British mystery fiction of the science of pure detection. I use the word "science" advisedly, for the man behind J. J. Connington, Alfred Walter Stewart, was an esteemed Scottish-born scientist who held the Chair of Chemistry at Queens University, Belfast for twenty-five years, from 1919 until his retirement in 1944. A "small, unassuming, moustached polymath," Stewart was "a strikingly effective lecturer with an excellent sense of humor, fertile imagination, and fantastically retentive memory," qualities that also served him well in his fiction. During roughly this period, the busy Professor Stewart found time to author a remarkable apocalyptic science fiction tale, *Nordenholt's Million* (1923), a mainstream novel, *Almighty Gold* (1924), a collection of essays, *Alias J. J. Connington* (1947), and, between 1926 and 1947, twenty-four mysteries (all but one true tales of detection), many of them sterling examples of the Golden Age puzzle-oriented detective novel at its considerable best. "For those who ask first of all in a detective story for exact and mathematical accuracy in the construction of the plot," avowed a contemporary

London Daily Mail reviewer, "there is no author to equal the distinguished scientist who writes under the name of J. J. Connington."[1]

Alfred Stewart's background as a man of science is reflected in his fiction, not only in the impressive puzzle plot mechanics he devised for his mysteries but in his choices of themes and depictions of characters. Along with Stanley Nordenholt of *Nordenholt's Million*, a novel about a plutocrat's pitiless efforts to preserve a ruthlessly remolded remnant of human life after a global environmental calamity, the most notable character that Stewart created is Chief Constable Sir Clinton Driffield, the detective in seventeen of the twenty-four Connington crime novels. Driffield is one of crime fiction's most highhanded investigators, occasionally taking into his hands the functions of judge and jury as well as chief of police. Absent from Stewart's fiction is the hail-fellow-well-met quality found in John Street's works or the religious ethos suffusing those of Freeman Wills Crofts, not to mention the effervescent novel of manners style of the British Golden Age Crime Queens Dorothy L. Sayers, Margery Allingham, and Ngaio Marsh. Instead we see an often disdainful cynicism about the human animal and a marked admiration for detached supermen with superior intellects. For this reason, reading a Connington novel can be a challenging experience for modern readers inculcated in gentler social beliefs. Yet Alfred Stewart produced a classic apocalyptic science fiction tale in *Nordenholt's Million* (justly dubbed "exciting and terrifying reading" by the *Spectator*), as well as superb detective novels boasting well-wrought puzzles, bracing characterization, and an occasional leavening of dry humor. Not long after Stewart's death in 1947, the Connington novels fell entirely out of print. The recent embrace of Stewart's fiction in recent publishing is a welcome

---

[1] For more on Street, Crofts and particularly Stewart, see Curtis Evans, *Masters of the "Humdrum" Mystery: Cecil John Charles Street, Freeman Wills Crofts, Alfred Walter Stewart and the British Detective Novel, 1920-1961* (Jefferson, NC: McFarland, 2012). On the academic career of Alfred Walter Stewart, see his entry in *Oxford Dictionary of National Biography* (London and New York: Oxford University Press, 2004), vol. 52, 627-628.

event indeed, correcting as it does over sixty years of underserved neglect of an accomplished genre writer.

Born in Glasgow on September 5, 1880, Alfred Stewart had significant exposure to religion in his earlier life. His father was William Stewart, longtime Professor of Divinity and Biblical Criticism at Glasgow University, and he married Lily Coats, a daughter of the Reverend Jervis Coats and member of one of Scotland's preeminent Baptist families. Religious sensibility is entirely absent from the Connington corpus, however. A confirmed secularist, Stewart once referred to one of his wife's brothers, the Reverend William Holms Coats (1881-1954), principal of the Scottish Baptist College, as his "mental and spiritual antithesis," bemusedly adding: "It's quite an education to see what one would look like if one were turned into one's mirror-image."

Stewart's J. J. Connington pseudonym was derived from a nineteenth-century Oxford Professor of Latin and translator of Horace, indicating that Stewart's literary interests lay not in pietistic writing but rather in the pre-Christian classics ("I prefer the *Odyssey* to *Paradise Lost*," the author once avowed). Possessing an inquisitive and expansive mind, Stewart was in fact an uncommonly well-read individual, freely ranging over a variety of literary genres. His deep immersion in French literature and supernatural horror fiction, for example, is documented in his lively correspondence with the noted horologist Rupert Thomas Gould.[2]

It thus is not surprising that in the 1920s the intellectually restless Stewart, having achieved a distinguished middle age as a highly regarded man of science, decided to apply his creative energy to a new endeavor, the writing of fiction. After several years he settled,

---

[2] The Gould-Stewart correspondence is discussed in considerable detail in *Masters of the "Humdrum" Mystery*. For more on the life of the fascinating Rupert Thomas Gould, see Jonathan Betts, *Time Restored: The Harrison Timekeepers and R. T. Gould, the Man Who Knew (Almost) Everything* (London and New York: Oxford University Press, 2006) and the British film *Longitude* (2000), which details Gould's restoration of the marine chronometers built by in the eighteenth-century by the clockmaker John Harrison.

like other gifted men and women of his generation, on the wildly popular mystery genre. Stewart was modest about his accomplishments in this particular field of light fiction, telling Rupert Gould later in life that "I write these things [what Stewart called tec yarns] because they amuse me in parts when I am putting them together and because they are the only writings of mine that the public will look at. Also, in a minor degree, because I like to think some people get pleasure out of them." No doubt Stewart's single most impressive literary accomplishment is *Nordenholt's Million*, yet in their time the two dozen J. J. Connington mysteries did indeed give readers in Great Britain, the United States, and other countries much diversionary reading pleasure. Today these works constitute an estimable addition to British crime fiction.

After his 'prentice pastiche mystery, *Death at Swaythling Court* (1926), a rural English country house tale set in the highly traditional village of Fernhurst Parva, Stewart published another, superior country house affair, *The Dangerfield Talisman* (1926), a novel about the baffling theft of a precious family heirloom, an ancient, jewel-encrusted armlet. This clever murderless tale, which likely is the one that the author told Rupert Gould he wrote in under six weeks, was praised in *The Bookman* as "continuously exciting and interesting" and in the *New York Times Book Review* as "ingeniously fitted together and, what is more, written with a deal of real literary charm." Despite its virtues, however, *The Dangerfield Talisman* is not fully characteristic of mature Connington detective fiction. The author needed a memorable series sleuth, more representative of his own forceful personality.

It was the next year, 1927, that saw "J. J. Connington" make his break to the front of the murdermongerer's pack with a third country house mystery, *Murder in the Maze*, wherein debuted as the author's great series detective the assertive and acerbic Sir Clinton Driffield, along with Sir Clinton's neighbor and "Watson," the more genial (if much less astute) Squire Wendover. In this much praised novel, Stewart's detective duo confronts some truly diabolical doings, including slayings by means of curare-tipped darts in the double-centered hedge maze at a country estate,

Whistlefield. No less a fan of the genre than T. S. Eliot praised *Murder in the Maze* for its construction ("we are provided early in the story with all the clues which guide the detective") and its liveliness ("The very idea of murder in a box hedge labyrinth does the author great credit, and he makes full use of its possibilities"). The delighted Eliot concluded that *Murder in the Maze* was "a really first-rate detective story." For his part, the critic H. C. Harwood declared in *The Outlook* that with the publication of *Murder in the Maze* Connington demanded and deserved "comparison with the masters." "Buy, borrow, or—anyhow get hold of it," he amusingly advised. Two decades later, in his 1946 critical essay "The Grandest Game in the World," the great locked room detective novelist John Dickson Carr echoed Eliot's assessment of the novel's virtuoso setting, writing: "These 1920s . . . thronged with sheer brains. What would be one of the best possible settings for violent death? J. J. Connington found the answer, with *Murder in the Maze*." Certainly in retrospect *Murder in the Maze* stands as one of the finest English country house mysteries of the 1920s, cleverly yet fairly clued, imaginatively detailed and often grimly suspenseful. As the great American true crime writer Edmund Lester Pearson noted in his review of *Murder in the Maze* in *The Outlook*, this Connington novel had everything that one could desire in a detective story: "A shrubbery maze, a hot day, and somebody potting at you with an air gun loaded with darts covered with a deadly South-American arrow-poison—*there* is a situation to wheedle two dollars out of anybody's pocket."[3]

Staying with what had for him worked so well, Stewart the same year produced yet another country house mystery, *Tragedy at Ravensthorpe*, an ingenious tale of murders and thefts at the ancestral home of the Chacewaters, old family friends of Sir Clinton Driffield. There is much clever matter in *Ravensthorpe*. Especially fascinating is the authors inspired integration of faerie folklore into his plot. Stewart, who had a lifelong—though skeptical—interest

---

[3] Potential purchasers of *Murder in the Maze* should keep in mind that $2 in 1927 is worth over $26 today!

in paranormal phenomena, probably was inspired in this instance by the recent hubbub over the Cottingley Faeries photographs that in the early 1920s had famously duped, among other individuals, Arthur Conan Doyle.[4] As with *Murder in the Maze*, critics raved this new Connington mystery. In the *Spectator*, for example, a reviewer hailed *Tragedy at Ravensthorpe* in the strongest terms, declaring of the novel: "This is more than a good detective tale. Alike in plot, characterization, and literary style, it is a work of art."

In 1928 there appeared two additional Sir Clinton Driffield detective novels, *Mystery at Lynden Sands* and *The Case with Nine Solutions*. Once again there was great praise for the latest Conningtons. H. C. Harwood, a critic who, as we have seen, had so much admired *Murder in the Maze*, opined of *Mystery at Lynden Sands* that it "may just fail of being the detective story of the century," while in the United States author and book reviewer Frederic F. Van de Water expressed nearly as high an opinion of *The Case with Nine Solutions*. "This book is a thoroughbred of a distinguished lineage that runs back to 'The Gold Bug' of [Edgar Allan] Poe," he avowed. "It represents the highest type of detective fiction." In both of these Connington novels, Stewart moved away from his customary country house milieu, setting *Lynden Sands* at a fashionable beach resort and *Nine Solutions* at a scientific

---

[4] In a 1920 article in *The Strand Magazine* Arthur Conan Doyle endorsed as real prank photographs of purported fairies taken by two English girls in the garden of a house in the village of Cottingley. In the aftermath of the Great War Doyle had become a fervent believer in Spiritualism and other paranormal phenomena. Especially embarrassing to Doyle's admirers today, Doyle also published *The Coming of the Faeries* (1922), wherein he argued that these mystical creatures genuinely existed. "When the spirits came in, the common sense oozed out," Stewart once wrote bluntly to his friend Rupert Gould of the creator of Sherlock Holmes. Like Gould, however, Stewart had an intense interest in the subject of the Loch Ness Monster, believing that he, his wife and daughter had cited a large marine creature of some sort in Loch Ness in 1935. A year earlier Gould had authored *The Loch Ness Monster and Others*, and it was this book which led Stewart, after he made his "Nessie" sighting, to initiate correspondence with Gould.

research institute. *Nine Solutions* is of particular interest today, I think, for its relatively frank sexual subject matter and its modern urban setting among science professionals, which rather resembles the locales found in P. D. James' classic detective novels *A Mind to Murder* (1963) and *Shroud for a Nightingale* (1971).

By the end of the decade of the 1920s, the critical reputation of "J. J. Connington" had achieved enviable heights indeed. At this time Stewart became one of the charter members of the Detection Club, an assemblage of the finest writers of British detective fiction that included, among other distinguished individuals, Agatha Christie, Dorothy L. Sayers and G. K. Chesterton. Certainly Victor Gollancz, the British publisher of the J. J. Connington mysteries, did not stint praise for the author, informing readers that "J. J. Connington is now established as, in the opinion of many, the greatest living master of the story of pure detection. He is one of those who, discarding all the superfluities, has made of deductive fiction a genuine minor art, with its own laws and its own conventions."

Such warm praise for J. J. Connington makes it all the more surprising that at this juncture the esteemed author tinkered with his successful formula by dispensing with his original series detective. In the fifth Clinton Driffield detective novel, *Nemesis at Raynham Parva* (1929), Alfred Walter Stewart, rather like Arthur Conan Doyle before him, seemed with a dramatic dénouement to have devised his popular series detective's permanent exit from the fictional stage (read it and see for yourself). The next two Connington detective novels, *The Eye in the Museum* (1929) and *The Two Tickets Puzzle* (1930), have a different series detective, Superintendent Ross, a rather dull dog of a policeman. While both these mysteries are competently done—the railway material in *The Two Tickets Puzzle* is particularly effective and should have appeal today—the presence of Sir Clinton Driffield (no superfluity he!) is missed.

Probably Stewart detected that the public minded the absence of the brilliant and biting Sir Clinton, for the Chief Constable—accompanied, naturally, by his friend Squire Wendover—triumphantly returned in 1931 in *The Boathouse Riddle*, another well-constructed criminous country house affair. Later in the year came *The Sweepstake Murders*, which boasts the perennially popular tontine multiple murder plot, in this case a rapid succession of puzzling suspicious deaths afflicting the members of a sweepstake syndicate that has just won nearly 250,000 pounds.[5] Adding piquancy to this plot is the fact that Wendover is one of the imperiled syndicate members. Altogether the novel is, as the late Jacques Barzun and his colleague Wendell Hertig Taylor put it in *A Catalogue of Crime* (1971/1989), their magisterial survey of detective fiction, "one of Connington's best conceptions."

Stewart's productivity as a fiction writer slowed in the 1930s, so that, barring the year 1938, at most only one new Connington appeared annually (because of the onset of serious health maladies, Stewart was unable to publish any Connington novel in 1936). However, in 1932 Stewart produced one of the best Connington mysteries, *The Castleford Conundrum*. A classic country house detective novel, Castleford introduces to readers Stewart's most delightfully unpleasant set of greedy relations and one of his most deserving murderees, Winifred Castleford. Stewart also fashions a wonderfully rich puzzle plot, full of meaty material clues for the reader's delectation. *Castleford* presented critics with no conundrum over its quality. "In *The Castleford Conundrum* Mr. Connington goes to work like an accomplished chess-player. The moves in the games his detectives are called on to play are a delight to watch," raved the reviewer for the *Sunday Times*, adding that "the clues would have rejoiced Mr. Holmes' heart." For its part, the *Spectator* concurred in the *Sunday Times*' assessment of the novel's masterfully-constructed plot: "Few detective stories show such sound reasoning as that by which the Chief Constable brings the crime home to the culprit." Additionally, E. C. Bentley, much

---

[5] A tontine is a financial arrangement wherein shareowners in a common fund receive annuities that increase in value with the death of each participant, with the entire amount of the fund going to the last survivor. The impetus that the tontine provided to the deadly creative imaginations of Golden Age mystery writers should be sufficiently obvious.

admired himself as the author of the landmark detective novel *Trent's Last Case*, took time to praise Connington's purely literary virtues, noting: "Mr. Connington has never written better, or drawn characters more full of life."

With *Tom Tiddler's Island* in 1933 Stewart produced a different sort of Connington, a criminal gang mystery in the rather more breathless style of such hugely popular English thriller writers as Sapper, Sax Rohmer, John Buchan and Edgar Wallace (in violation of the strict detective fiction rules of Ronald Knox, there is even a secret passage in the novel). Detailing the startling discoveries made by a newlywed couple honeymooning on a remote Scottish island, *Tom Tiddler's Island* is an atmospheric and entertaining tale, though it is not as mentally stimulating for armchair sleuths as Stewart's true detective novels. The title, incidentally, refers to an ancient British children's game, "Tom Tiddler's Ground," in which one child tries to hold a height against other children.

After his fictional Scottish excursion into thrillerdom, Stewart returned the next year to his English country house roots with *The Ha-Ha Case* (1934), his last masterwork in this classic mystery setting. (For elucidation of non-British readers, a ha-ha is a sunken wall, placed so to delineate property boundaries while not obstructing views.) Although *The Ha-Ha Case* is not set in Scotland, Stewart drew inspiration for the novel from a notorious Scottish true crime, the 1893 Ardlamont murder case. From the facts of the Ardlamont affair Stewart drew several of the key characters in *The Ha-Ha Case*, as well as the circumstances of the novel's murder (a shooting "accident" while hunting), though he added complications that take the tale in a new direction.[6]

---

[6] At Ardlamont, a large country estate in Argyll, Cecil Hambrough died from a gunshot wound while hunting. Cecil's tutor, Alfred John Monson, and another man, both of whom were out hunting with Cecil, claimed that Cecil had accidentally shot himself; but Monson was arrested and tried for Cecil's murder. The verdict delivered was "not proven," but Monson was then—and is today—considered almost certainly to have been guilty of the murder. On the Ardlamont case, see William Roughead, *Classic Crimes* (1951; repr., New York: New York Review Books Classics, 2000), 378-464.

In newspaper reviews both Dorothy L. Sayers and "Francis Iles" (crime novelist Anthony Berkeley Cox) highly praised this latest mystery by "The Clever Mr. Connington," as he was now dubbed on book jackets by his new English publisher, Hodder and Stoughton. Sayers particularly noted the effective characterization in *The Ha-Ha Case*: "There is no need to say that Mr. Connington has given us a sound and interesting plot, very carefully and ingeniously worked out. In addition, there are the three portraits of the three brothers, cleverly and rather subtly characterised, of the [governess], and of Inspector Hinton, whose admirable qualities are counteracted by that besetting sin of the man who has made his own way: a jealousy of delegating responsibility." The reviewer for the *Times Literary Supplement* detected signs that the sardonic Sir Clinton Driffield had begun mellowing with age: "Those who have never really liked Sir Clinton's perhaps excessively soldierly manner will be surprised to find that he makes his discovery not only by the pure light of intelligence, but partly as a reward for amiability and tact, qualities in which the Inspector [Hinton] was strikingly deficient." This is true enough, although the classic Sir Clinton emerges a number of times in the novel, as in his subtly sarcastic recurrent backhanded praise of Inspector Hinton: "He writes a first class report."

Clinton Driffield returned the next year in the detective novel *In Whose Dim Shadow* (1935), a tale set in a recently erected English suburb, the denizens of which seem to have committed an impressive number of indiscretions, including sexual ones. The intriguing title of the British edition of the novel is drawn from a poem by the British historian Thomas Babington Macaulay: "Those trees in whose dim shadow/The ghastly priest doth reign/The priest who slew the slayer/And shall himself be slain." Stewart's puzzle plot in *In Whose Dim Shadow* is well-clued and compelling, the kicker of a closing paragraph is a classic of its kind and, additionally, the author paints some excellent character portraits. I fully concur in the *Sunday Times* assessment of the tale: "Quiet domestic murder, full of the neatest detective points. . . . These [characters] are not the detective's stock figures, but fully realised human beings."[7]

Uncharacteristically for Stewart, nearly twenty months elapsed between the publication of *In Whose Dim Shadow* and his next book, *A Minor Operation* (1937). The reason for the author's delay in production was the onset in 1935-36 of the afflictions of cataracts and heart disease (Stewart ultimately succumbed to heart disease in 1947). Despite the grave health complications that beset him at this time, Stewart in late 1936 was able to complete *A Minor Operation*, a first-rate Clinton Driffield story of murder and a most baffling disappearance. A *Times Literary Supplement* reviewer found that *A Minor Operation* treated the reader "to exactly the right mixture of mystification and clue" and that, in addition to its impressive construction, the novel boasted "character-drawing above the average" for a detective novel.

Alfred Stewart's final eight mysteries, which appeared between 1938 and 1947, the year of the author's death, are, on the whole, a somewhat weaker group of tales than the sixteen that appeared between 1926 and 1937, yet they are not without interest. In 1938 Stewart for the last time managed to publish two detective novels, *Truth Comes Limping* and *For Murder Will Speak*. The latter tale is much the superior of the two, having an interesting suburban setting and a bevy of female characters found to have motives when a contemptible philandering businessman meets with foul play. Sexual neurosis plays a major role in *For Murder Will Speak*, the

---

[7] For the genesis of the title, see Macaulay's "The Battle of the Lake Regillus," from his narrative poem collection *Lays of Ancient Rome*. In this poem Macaulay alludes to the ancient cult of Diana Nemorensis, which elevated its priests through trial by combat. Study of the practices of the Diana Nemorensis cult influenced Sir James George Frazer's cultural interpretation of religion in his most renowned work, *The Golden Bough: A Study in Magic and Religion*. As with *Tom Tiddler's Island* and *The Ha-Ha Case* the title *In Whose Dim Shadow* proved too esoteric for Connington's American publishers, Little, Brown and Co., who altered it to the more prosaic *The Tau Cross Mystery*.

ever-thorough Stewart obviously having made a study of the subject when writing the novel. The somewhat squeamish reviewer for *Scribner's Magazine* considered the subject matter of *For Murder Will Speak* "rather unsavory at times," yet this individual conceded that the novel nevertheless made "first-class reading for those who enjoy a good puzzle intricately worked out." "Judge Lynch" in the *Saturday Review* apparently had no such moral reservations about the latest Clinton Driffield murder case, avowing simply of the novel: "They don't come any better."

Over the next couple years Stewart again sent Sir Clinton Driffield temporarily packing, replacing him with a new series detective, a brash radio personality named Mark Brand, in *The Counsellor* (1939) and *The Four Defences* (1940). The better of these two novels is *The Four Defences*, which Stewart based on another notorious British true crime case, the Alfred Rouse blazing car murder. (Rouse is believed to have fabricated his death by murdering an unknown man, placing the dead man's body in his car and setting the car on fire, in the hope that the murdered man's body would be taken for his.) Though admittedly a thinly characterized academic exercise in ratiocination, Stewart's *Four Defences* surely is also one of the most complexly plotted Golden Age detective novels ever written and should delight devotees of classical detection. Taking the Rouse blazing car affair as his theme, Stewart composes from it a stunning set of diabolically ingenious criminal variations. "This is in the cold-blooded category which . . . excites a crossword puzzle kind of interest," the reviewer for the *Times Literary Supplement* acutely noted of the novel. "Nothing in the Rouse case would prepare you for these complications upon complications. . . . What they prove is that Mr. Connington has the power of penetrating into the puzzle-corner of the brain. He leaves it dazedly wondering whether in the records of actual crime there can be any dark deed to equal this in its planned convolutions."

Sir Clinton Driffield returned to action in the remaining four detective novels in the Connington oeuvre, *The Twenty-One Clues* (1941), *No Past Is Dead* (1942), *Jack-in-the-Box* (1944) and *Commonsense Is All You Need* (1947), all of which were written as

Stewart's heart disease steadily worsened and reflect to some extent his diminishing physical and mental energy. Although *The Twenty-One Clues* was inspired by the notorious Hall-Mills double murder case—probably the most publicized murder case in the United States in the 1920s—and the American critic Anthony Boucher commended *Jack-in-the-Box*, I believe the best of these later mysteries is *No Past Is Dead*, which Stewart partly based on a bizarre French true crime affair, the 1891 Achet-Lepine murder case.[8] Besides providing an interesting background for the tale, the ailing author managed some virtuoso plot twists, of the sort most associated today with that ingenious Golden Age Queen of Crime, Agatha Christie.

What Stewart with characteristic bluntness referred to as "my complete crack-up" forced his retirement from Queen's University in 1944. "I am afraid," Stewart wrote a friend, the chemist and forensic scientist F. Gerald Tryhorn, in August, 1946, eleven months before his death, "that I shall never be much use again. Very stupidly, I tried for a session to combine a full course of lecturing with angina pectoris; and ended up by establishing that the two are immiscible." He added that since retiring in 1944, he had been physically "limited to my house, since even a fifty-yard crawl brings on the usual cramps." Stewart completed his essay collection and a final novel before he died at his study desk in his Belfast home on July 1, 1947, at the age of sixty-six. When death came to the author he was busy at work, writing.

More than six decades after Alfred Walter Stewart's death, his "J. J. Connington" fiction again is available to a wider audience of classic mystery fans, rather than strictly limited to a select company of rare book collectors with deep pockets. This is fitting for an individual who was one of the finest writers of British genre fiction between the two world wars. "Heaven forfend that you should imagine I take myself for anything out of the common in

---

[8] Stewart analyzed the Achet-Lepine case in detail in "The Mystery of Chantelle," one of the best essays in his 1947 collection, *Alias J. J. Connington*.

the tec yarn stuff," Stewart once self-deprecatingly declared in a letter to Rupert Gould. Yet, as contemporary critics recognized, as a writer of detective and science fiction Stewart indeed was something out of the common. Now more modern readers can find this out for themselves. They have much good sleuthing in store.

ALIAS J. J. CONNINGTON

PREFACE

HOW DOES IT FEEL to pass under an alias? I ought to know, inasmuch as I have appeared under a brace of names for almost a quarter of a century. During that time, I kept two sides of my work independent of each other by publishing my scientific books and research papers under my own name, whilst any fiction of mine was issued under the pseudonym J. J. Connington.

But before one can pass under an alias, one must seek out something suitable to disguise one's identity, and how is that to be found? One way is to go to the classics. Junius, Peter Pindar, and Democritus Junior all served good purpose in their day; but this sort of thing has now fallen into the hands of political pamphleteers, and is best avoided by the rest of us.

Another method is to turn one's own name into an anagram, as Rabelais did when he posed as Maître Alcofribas Nasier, and Arouet le jeune when he blossomed out into Voltaire. But my own name does not lend itself to anagrammatisation, so I had to seek elsewhere.

Actually, I borrowed the surname from the translator of Horace, added an extra "n" for good measure, and tacked on J. J. as being a pair of easily-recalled initials. But then the Library of Congress in the United States took a hand in the matter; and in order to satisfy their curiosity I had to invent two Christian names to fit the double J.

Aliases are not confined to real life. There is, for example, what Forster termed "the Datchery assumption" in *The Mystery of*

Edwin Drood. What was the real name of Datchery, that "single buffer, of an easy temper, living idly on his means"? No one knows; no one will ever know, though plenty of people have tried to guess it.

R. L. Stevenson, that Peter Pan in real life, had a penchant for supplying his characters with aliases. The Master of Ballantrae figures as Mr. Bally through a large part of his history. The prisoner of war in *St. Ives* is, naturally enough, a perfect chameleon in the matter of *noms de guerre*. In *The Black Arrow* we find Ellis Duckworth sheltering behind John Amend-all and Joanna Sedley disguised as John Matcham. Whether or not Mr. Hyde was, properly speaking, an alias for Dr. Jekyll, seems a moot point which I leave to the judgment of my readers. But in *The New Arabian Nights*, Prince Florizel of Bohemia is content to pass as Mr. Godall, while Colonel Geraldine plays Major Hammersmith. The trio of scoundrels in *The Ebb-Tide* all have purser's names; whilst in *The Wrecker* the whole crew of the *Currency Lass* merge their identities in those of the *Flying Scud's* complement.

It is in *The Wrong Box*, however, that the thing attains its full florescence. John Finsbury, a person of no originality, was content to pose as the Great Vance, a part which he was singularly ill-fitted to sustain, as the result showed. The unfortunate little drawing-master, William Dent Pitman, found himself unwillingly transmuted into Ezra Thomas, a wealthy manufacturer of india-rubber overshoes and pirated Broadwood pianos, with cotton-mills in Tallahassee, tobacco-factories in Richmond, Va., "and many, many, many others". Michael Finsbury chose to become in turn John Dickson of Ballarat and one Appleby. Gideon Forsyth, after publishing *Who Put Back the Clock?* under the initials E.H.B., tried to conceal himself under the guise of the Maestro Jimson, but so unsuccessfully that even that gentle soul Julia Hazeltine condemned him as "a mere impostor" as soon as she set eyes on his musical composition *Orange Pekoe*—which turned out to be merely *Tommy Make Room for Your Uncle*, so that even music passes under an alias. And even this lengthy catalogue omits the confusion of identity between Joseph Finsbury and the late patient of Sir Faraday Bond. Truly an amazing mix-up.

In the course of his holiday, Michael Finsbury emitted the aphorism: "One drunken man, excellent business—two drunken men, all my eye." Much the same might be said about the plethora of aliases in *The Wrong Box*. Evidently, with Stevenson, the thing was an obsession which clung to him throughout his career. I can recall no other writer who scattered pseudonyms so freely among his characters in volume after volume.

But for catholicity in aliases, the palm must be awarded to Lewis Carroll's baker in *The Hunting of the Snark*.

"He would answer to 'Hi!' or to any loud cry."

This opens vistas extending "ever so far" as Euclid says when speaking of parallel lines.

In my own case, the difficulty of concealing one's real identity behind an alias was brought home to me at the very start, when I wrote a pseudo-scientific thriller called *Nordenholt's Million*. An astute friend of mine came across the book by chance, and noted three things. In the first place, it dealt with chemical subjects in which he knew I was interested, such as atomic energy and the nitrogen cycle in Nature. In the second place, the book's topography was mainly concerned with the Clyde Valley, Glasgow University, and the district of London around University College, all of which he knew to be specially familiar to me. Finally, the volume was dedicated "To J.N.C.", the initials of my old chief, Professor John Norman Collie, F.R.S., about whom I shall have something to say later in this volume. Basing himself on these three coincidences, my friend wrote to me, taxing me with the authorship of *Nordenholt's Million*, and I had to confess that his inferences had led him to the right door. So flimsy is an alias.

My Publishers have asked me if there is any connection between my work as a scientific investigator on the one hand and my detective-story writing on the other. There is not the slightest parallelism between these two lines, except that in both a logical mind is required.

In scientific research, the inquirer plays the part of a detective in real life. It is his business to seek for relevant clues; to push red herrings off the scent; to free himself from any preconceived ideas about the solution, if they conflict with the facts which he is collecting, often at the cost of wearisome labour; and, finally, to present a mass of detailed evidence in such a way that it will carry conviction.

On the other hand, the author of a detective story begins with the solution already in his pocket, and his initial task is to invent a "chain of evidence" which leads inevitably to that solution and to no other. But this, in itself, is hardly sufficient, for the good detective story is really a duplex affair. Superficially, it presents a series of events as seen by Dr. Watson or his deputy; below that lies concealed the tale of what actually happened; and the skill of the author is to be gauged from the manner in which he can blend these two themes in such a way that the reader, after perusing the story, can go back and start afresh, punctuating his second reading with the ejaculation: "What a fool I was not to have noticed that point the first time!" Any well-contrived detective story should admit of at least two perusals if its full interest is to be extracted from it. When, in addition to this, the author is able to make the plot appear to arise from the characters—as Mr. A. E. W. Mason did in *At the Villa Rose*—then his work touches the high-water mark of detective story writing and constitutes a rarely-achieved feat which fills less skilful authors with mingled admiration and envy.

Evidently, then, there is no resemblance between the scientific investigator, working from details towards a solution, and the detective-story writer, starting from his preconceived solution and working back to the details of his plot. To find an analogy, we must seek elsewhere, in another field of art.

The closest likeness to the writing of a detective story is to be found in the composition of a chess-problem. In both cases, the constructor begins at the end and works backward. The chess expert, having hit upon his checkmate position, has then to devise moves leading up to it, and has to place on the board a pawn here or a piece on some other square, in order to block certain moves which might otherwise be made. In the same way, the writer of a detective story has to invent various characters and episodes which limit the possibilities in his imagined course of events leading up to his solution. In both fields, it is permissible to introduce in moderation a supply of red herrings which serve merely to distract attention and play no essential part in the puzzle. And in both chess-problem and detective story there is a "key-move" or a key-episode, which must be fixed upon by the would-be solver if he is to get immediately upon the track of the solution.

But enough of aliases and detective stories.

This is an "escape" book, intended to divert the reader for a short time from the troubles which daily engross us. It deals with a number of topics: a murder mystery in real life; the germs of the plots of some well-known novels; the curious problem of the Dumbuck crannog; the psychological illusions produced by certain drugs; various methods of getting "easy money"; the value of an honest witness's evidence; the adventurous careers of the translators of the *Arabian Nights*; the discovery of the black swan; the rather gruesome history of a millionaire mystic; the Loch Arkaig treasure; the strange affair of the Long Man on Ben Macdhui; my own casual encounter with that curious phenomenon which goes by the name of the Loch Ness Monster; and other subjects as well.

Part of the material comes from my experiences as a writer, a professor, a technical adviser, and a scientific researcher; but the reader may rely on the assurance that there is no likelihood of him finding himself out of his depth, even if he has no acquaintance with science. Other portions of the volume are the results of a life-long predilection for browsing off the beaten track in my reading. These parts, I hope, may act as finger-posts to direct my readers deeper into some fascinating fields which I have here treated cursorily for lack of elbow-room.

My aim has been to provide a wide variety of topics, so that there may be "something for everybody"; and I may fittingly conclude, in the words of an older writer: "Choose that which pleaseth thee best. Not to detain thee longer, farewell; and when thou hast considered thy purchase, may'st thou say, that the price of it was but a charity to thyself, not ill spent."

A. W. Stewart.

I
THE MYSTERY OF CHANTELLE

IN THESE DAYS when the art and science of murder has become the delectation of refined intellects, it seems surprising that the mystery of Chantelle has attracted so little attention. There is no obscurity about the crime itself: a determined assassination wherein revolver and knife both played their parts and the corpse was left exposed for anyone to see. The real puzzle lies in the course of events which led up to the murder, for, despite the efforts of the French authorities, uncertainty clouds the whole history of the affair.

The curious financial transactions which preceded the murder are capable of various interpretations, none of which reflects much credit upon the persons involved. Then, with the lawyer's nocturnal visit to his client's house, ambiguousness deepens. Did Lépine go there merely to reclaim his money? Or, having got an attractive woman into his clutches by his loans, was he bent on an errand of ill-calculated gallantry? Or, again, did Mme. Achet draw him into an ambuscade in the hope of ridding herself once for all of the monetary troubles which beset her? Was she alone, or was there an accomplice lying in wait to play his part in the murder? If so, was he one of Mme. Achet's lovers or was he some local ruffian seduced by the promise of a cash reward? These are problems on which it is fruitless to speculate, yet interesting to muse upon.

Other points suggest themselves in a study of the trial. There is the detachment shown by Mme. Achet during her ordeal, an aloofness curiously reminiscent of the demeanour of Madeleine

Smith in a like situation. There is the complete incuriosity shown by some of the witnesses at the moment of the tragedy. One of them heard the noise of four revolver shots, followed by a loud cry; he calmly closed his window and went to bed. Others showed hardly more interest. Again, there was the familiar figure of the man who confessed that he was the murderer, but who later was found to have no more connection with the affair than the Man in the Moon. As it was a French case, politics were almost inevitably dragged into the arena—the politics of a tiny town with all its little jealousies—and they occupied a certain amount of the Court's time, without throwing the slightest light upon the crime.

The difference between British and French procedure comes out when it appears that the accused was required not only to defend herself against a charge of murder, but *simultaneously* to fight a civil action to protect her pocket; and, to English ideas, the last surprise is added when the evidence in the civil suit is lumped in with that of the criminal trial and heard in detail before the verdict is delivered.

Finally—a touch which Poe would have relished—the shadow of the madhouse looms in the background of the tragedy.

Mme. Achet, the daughter of a Parisian contractor, was married in 1877, at the age of nineteen, to a man twenty years older than herself who ran a crammer's establishment, preparing students for degrees. The financial position of the couple at the time of this marriage is indicated by the facts that Mme. Achet brought a dowry of 100,000 francs (£4,000) plus expectations from the succession to her grandmother's estate, estimated at about an equal figure, whilst her husband contributed to the common stock some 40,000 francs (£1,600) partly invested in land. Two children were born: a girl who died at the age of ten and a boy, Ali, who was seven years old at the time of the trial.

In 1880, Achet sold his cramming establishment and settled down in Chantelle, a little country town of about a thousand inhabitants, not far from Moulins. Here he launched out into various unfortunate speculations, with the result that when he died, semi-demented, in 1885, his wife's fortune was wasted away to such a degree that she was forced to borrow some 30,000 francs (£1,200) from the bank; and she found herself reduced to living on a net income of 3,000 francs (£120) per annum.

In a quiet little place like Chantelle, existence on this scale was possible; but it was not to the taste of Mme. Achet with her Parisian upbringing, her love of fine clothes, and her liking for trips to the capital, where she rented a room despite the fact that her parents' house was open to her. So debts accumulated; and, by 1890, her extravagant way of life had left her with an annual income of only 2,400 francs (£96) all told. This was hardly sufficient to maintain herself and her little boy, as well as to keep up the establishment at Chantelle: a pleasantly-situated, tile-roofed, green-shuttered house erected on the site of a medieval stronghold, where moss-grown walls enclosed the sloping grounds with terraces, lawns, rose-alleys, an orchard, a conservatory, and even—bizarre touch!—a Chinese kiosk.

In addition to the decline in her income, another change had come over Mme. Achet's affairs. During the lifetime of her husband, her conduct had been irreproachable, but her widowhood was not without consolations. There was nothing provincial about her; she had the type of good looks which sometimes goes with irregular features; one of her women-friends testified that she was "an angel of good-nature and charm"; and even at thirty-two she impressed one observer as having retained something of the irrepressible schoolgirl in her nature and manner. The actual number of her lovers is not material to the case; she admitted to four, but this was undoubtedly much below the true figure.

Strangely enough, though her money matters were conducted in the most haphazard fashion, she systematised her amatory affairs with a most unusual thoroughness. She was an indefatigable writer of love-letters; and she meticulously preserved those which she received, whole collections of them falling into the hands of the police eventually. She kept a register of those which she wrote day by day to her divers lovers, and another daily index of those which she received. An inspection of these lists showed that at times she wrote equally fervent letters to various admirers on the same day: a state of affairs which drew from the prosecutor the shocked comment that "Love, however, is an exclusive sentiment." One gathers that she had a very likeable personality, for despite her irregular life she was received in the best circles of Chantelle, though tiny places of that sort are not usually noted for their social charity.

Of Mme. Achet's numerous lovers, only two made any real appearance in the case, though a doctor in Paris was mentioned incidentally. In the first place, there was M. Delorme, a man of forty-two, who held a position in the Paris assurance company, *L'Abeille*. She received him from time to time in her discreet apartment in the Boulevard de Strasbourg; but her affair with him seems to have been a mere pastime with no ulterior object. More serious, perhaps, was M. Thaunié, a rich young gentleman of twenty-seven, a landed proprietor in the neighbourhood of Chantelle, who seems to have divided his time between hunting and looking after his estate. He may well have been marked down by Mme. Achet as a possible second husband. On the fatal night, M. Delorme was in Paris, whilst M. Thaunié had gone a-hunting in La Creuse in company with a lieutenant of the gendarmerie. Thus, each of these lovers of Mme. Achet had an irrefragible alibi.

At this point there comes into the story the enigmatic figure of Maître Lépine; and it is around his motives and conduct that the mystery gathers most thickly. He was a man of about fifty, married, with a grown-up son, and in his profession of provincial notary he seems to have been highly respected in the neighbourhood of Chantelle. Physically, he was rather repulsive owing to a large goitre. He lived only a short distance from the house of Mme. Achet, and after her husband's death he took charge of her legal and financial affairs. Though he seems to have been successful in raising money for her, the relations between lawyer and client do not appear to have gone altogether smoothly at times, as is shown by this extract from a letter written by Mme. Achet to one of her lovers:

> "It is childish to suppose that I have chosen this *stout goose*,* this uncivilised boor, as a confidant. It is he who plays the attacker, I defend myself. We are like two duellists. Things will soon be settled up."

Again, during the trial, a magistrate, M. Raillard, deponed that he had had some business dealings with Mme. Achet and that during the course of these he had received from her a letter in which she wrote:

> "I cannot understand M. Lépine's methods at all. On the one hand, he does me all the harm he can in my business affairs; but on the other hand he lends me all the money I need. One would think he coveted my house in Chantelle and wanted to force my hand."

By the end of June, 1890, Mme. Achet was in debt to Lépine to the extent of 3,800 francs (£232); and during the next three months she borrowed bit by bit another 800 francs (£32). Meanwhile, as the result of obscure and complicated transactions, she had been induced to sign an acknowledgement for 10,000 francs (£400), a document which appeared prominently in the later stages of the affair.

This brings the story up to October 5, 1890, and to one of its most mysterious episodes. On the evening of that day, and at an hour much later than one would expect a lawyer to pay a professional visit, Lépine called at Mme. Achet's house. What occurred during this interview is far from clear, but there is sufficient evidence to show that when he left the premises he "forgot" to take with him: (1) his pocket-book; (2) the acknowledgement for 10,000 francs; and (3) banknotes to the value of 7,200 francs (£288). During the next ten days, Lépine made various abortive efforts to

---

* The italicised words are in English in the original.

recover his 7,200 francs; and in the books which he kept he entered on the following day (October 6), "Handed to Mme. Achet, 7,200 francs."

During her preliminary examination, Mme. Achet was questioned on the subject. Her reply at that stage was noncommittal. "M. Lépine left his pocket-book with me of his own free will, and if I were to tell you the circumstances you would know the secret of a good many things."

In the actual trial, a little more light was thrown on the matter, by questions in Court.

> Q. "According to M. Lépine, he *forgot* this money at your house?"—A. "Not at all. He handed it to me. Something happened which I do not wish to describe." Q. "Come, come. We can't leave matters in darkness for ever. Will you answer?"—A. "One thing's certain: I didn't ask him for this money." Q. "Then what reason had he for giving it to you?"—A. "He was advancing that out of a loan in my name from one of my uncles. It was he who suggested this advance. He thought this would please me, leaving something else to be understood." Q. "What?"—A. (made after a long hesitation) "That he wouldn't ask for it back?" Q. Did he say he wouldn't ask for it back?"—A. "One doesn't say things like that in so many words." Q. "And what did you answer?"—A. "I broke off the interview." Q. "But you kept the money?"

Counsel for the defence was Maître Demange, later to be one of the barristers who defended Captain Dreyfus in the trial at Rennes. He evidently felt that a bad impression was left by the foregoing exchanges, and he requested the Judge to ask the accused "how it came about that the notary left the house so hurriedly that he forgot his pocket-book?" But Mme. Achet still fenced.

"I'll say nothing. But one thing's sure: he went away far from satisfied, and I felt the same."

But at this point the Court, weary of evasion, determined to have the facts if possible:

> Q. "Now, did he say in explicit terms that he would not ask for the return of the 7,000 francs provided you bestowed your favours on him?"—A. "If he didn't say it, he made me understand it."

Mme. Achet's account of this interview on October 5 is the only one extant; Lépine gave no detailed version of what happened, and this, in itself, may seem suggestive. We are, therefore, reduced to balancing against each other the narrative of the accused and the hypothesis suggested by the prosecutor, in order to see which is the more probable.

Baldly stated, the prosecution's theory amounts to this: that Lépine was merely an honest notary, perturbed by the financial situation and anxious to have it cleared up. But, in addition, as if casually, the prosecutor threw in a speculation: "One may ask if you are the sort of woman who is insulted by a man's desire or the sort of woman who is insulted when a man rejects your offers." In other words, Mme. Achet's evidence on one point was a complete inversion of the facts; and it was she, not Lépine, who proposed to make a settlement by payment in kind and not in cash.

This simple view of the case, however, leaves several questions unsolved. Why did Lépine choose a late hour for his business call? Why did he bring with him the written acknowledgement of the 10,000-franc debt and leave it behind? Why did he leave behind him also the 7,200 francs in notes without securing a receipt for them? And why, in entering up this last transaction in his books at home, did he mis-date it and so give the impression that the money was paid on the sixth of October instead of the fifth?

Turn now to the story told by Mme. Achet. According to her, Lépine came to her house and dangled before her a large financial bait in order to gain possession of her person. That tale gains a certain support from the letter quoted by M. Raillard during the trial, for in it one finds the hint that Lépine had an ulterior purpose in his financial operations. Countering the insinuation of the prosecution, the defence pointed out that Mme. Achet had not been what is now called a gold-digger. "She had loved her first paramour with an ideal love . . . She had loved a second, a third, without ever asking so much as a centime." One of her current lovers was M. Thaunié, the son of a millionaire, with a reputation for spending money like water; so that if money was needed, a source was not far to seek. In such circumstances, is it probable that she offered herself to a country notary, physically repulsive in his appearance by reason of his goitre? Or is it more likely that he offered her the money which she accepted and then snapped her fingers at him? Behind her chaste reluctance to go into details there is a suggestion that Lépine did not stop short of physical violence in an effort to gain his ends; and this is strengthened to some extent by the curious episode of the revolver, which is still to be described.

On the other hand, two points in her tale excite suspicion. In the first place, she alleged that Lépine presented her with the equivalent of 17,200 francs (£688) during the interview on October 5. That seems a fairly stiff price for even the most highly-enamoured country lawyer to pay for her favours. In the second place, her assertion about her uncle and the loan of 10,000 francs was a downright lie, for this uncle had died in the previous autumn and had left her nothing in his will.

One statement, however, may be taken as accurate: both parties to the conversation on October 5 were left highly dissatisfied at the end of it. Lépine's reaction may be gauged from the fact that he seems to have spent the next few days in clamouring for the return of his money. Mme. Achet's state of mind may be guessed from steps which she took shortly after the interview.

Some two years earlier, she had bought a revolver. Her story was that she meant to take it to Chantelle as a means of protection, since there were only women in her house; but actually, for part of the time, the weapon seems usually to have been left lying in her Paris apartment. On October 7, two days after her interview with Lépine, she wrote to her lover, Delorme, in Paris: "Please have my revolver repaired and send it on to me." And she enclosed in the letter a banknote for 1,000 francs (£40), "to cover expenses." She did not register this letter despite its value. Asked why she had neglected this precaution, she explained that she did not wish people in Chantelle to know that she had money, and, further, that the same reason prevented her from changing the note locally and induced her to send it to M. Delorme in Paris. M. Delorme, it appears, quite understood, for he at once sent her by registered post ten notes of 100 francs each. He did not, however, despatch the revolver immediately owing to some oversight; whereupon, not having received it, Mme. Achet telegraphed to him: "Please send back silk immediately." M. Delorme, evidently a perspicacious gentleman, penetrated this subterfuge immediately. I understood at once," he remarked, "that she meant the revolver, and that she did not want her affairs talked about in Chantelle. I sent her the revolver, just as I would have sent on an umbrella." Possibly to avoid gossip, M. Delorme labeled the packet: "Metal tubes." Thus, by October 14, Mme. Achet had her revolver at Chantelle.

Some stress was laid by the prosecution on the urgency and secrecy shown by Mme. Achet in this matter. There was, however, no secret about her actual possession of a revolver; some time previously, when the weapon happened to be at Chantelle, she had practised with it in her garden. As to her urgency in telegraphing for it, Mme. Achet explained that she had arranged to go for a trip with M. Thaunié, that they would have to leave in his carriage by dead of night to avoid attention, and that he had warned her that the roads were not safe. In the witness-box, Thaunié confirmed this story of a projected trip, though he put the date of it rather later than Mme. Achet did.

On the very day that the revolver came into her hands at Chantelle, Mme. Achet tried to arrange an appointment with Lépine at her house, at the hour of 10 p.m. It may be suggested that, with the weapon in her possession, she felt secure against any violence from Lépine during this interview. The prosecution, not unnaturally, saw it from a different standpoint: now that she had her revolver, she proposed to draw Lépine into a trap and put an end to the business. The notary, however, refused this appointment, as well as another at the same hour on the following night.

But by now, according to the view of the prosecution, Mme. Achet had been brought to bay. She owed Lépine 17,200 francs and she had no resources. Plainly she saw disaster impending over her, and in desperation she sought for some path, any path, out of her entanglements. She wrote to Lépine, demanding the return of "certain papers"—the nature of which was never disclosed. She begged for a scholarship to provide for the education of young Ali. She made plans for leaving Chantelle. She even wrote to her lover, the Paris doctor, offering to "associate her life with his"—and the doctor ignored her note. A phrase in a letter to her paramour Delorme throws a lurid flash of illumination on her agitation at this moment: "I am dancing on a spring-board; who knows where I shall land!"

One rubs one's eyes. What of Thaunié, who seems to have been lavish with his money? He was fond of her. A matter of six or seven hundred pounds would hardly have broken the son of a millionaire. Undoubtedly, Mme. Achet was no gold-digger; she had not tried to make money out of her love-affairs. But she might well have been forgiven if she had sought Thaunié's financial aid in this plight. Yet, so far as the evidence goes, she seems to have completely forgotten him in these final straits.

For final they were. Lépine had sent her an ultimatum saying that he was "determined to have the whole matter settled up" on October 17. Mme. Achet replied, telling him to call on her about five o'clock and to bring with him "all the necessary documents"—whatever that phrase implied.

But when Lépine presented himself about 5 p.m., he was told that Mme. Achet had her dressmaker in the house and could not give him any time. He proposed to come later. Here is Mme. Achet's evidence on the point:

Q. "At what time did he propose to come back?"—A. "The same evening, but I told him I could not receive him at such a late hour." Q. "But you had sometimes received him in the evening?"—A. "Sometimes, but I was done with that. 'No,' I said, 'come back tomorrow; I am spending the evening with my sister, Mme. Lamotte, who lives in Chantelle.'" Q. "But did you not tell him to come back at half-past ten, after your return home?"—A. "No one settles an account at ten in the evening."

For information about Lépine's last hours, it is necessary to turn to the evidence given by his widow and his son at the trial. During the evening, they played cards, and Lépine seemed perfectly cool. He made no secret of his coming visit to Mme. Achet; and his wife was perturbed at the lateness of the hour. She pointed out to him that ten o'clock was no time for conducting business. Lépine answered that Mme. Achet did not wish her sister to know that she was doing business with him. This seemed curious, since Lépine had, for a good while, been collecting Mme. Achet's income for her and paying debts on her behalf; a process which could hardly have been private.

Now, as it chanced, the previous year had seen a notorious murder in Paris; a bailiff, Gouffé, had been lured to the house of a demi-mondaine, Gabrielle Bompard, and there murdered by her accomplice, Michael Eyraud. The case had made a sensation in France, and its details were known even in the provinces. Mme. Lépine suffered from presentiments; and on this evening she had a strong one. She told her husband that if he went to Mme. Achet's, he would come by the same fate as Gouffé. So acute was her misgiving that, at the moment when Lépine left his house at ten o'clock, she called to him from her bed to take his revolver with him, which he did. He took with him also the accounts relating to Mme. Achet's affairs and, presumably, the mysterious "documents" as well. Apparently he did not expect his absence to be a long one, for he left his lamp burning.

Meanwhile, Mme. Achet spent the evening with her sister, Mme. Lamotte, who was married to the Chantelle druggist. About half-past nine her maid, Françoise Mosnier, a stolid peasant-woman of forty-five, called at M. Lamotte's in order to bear her company home, as seems to have been the customary practice. Mme. Achet wished to stay longer and have some music, but M. Lamotte was sleepy and turned her out.

There were two entrances to the grounds of Mme. Achet's house; an iron gate and a small side gate. The two women went in by the iron gate, which the maid locked after them, handing the key to her mistress. Françoise then went up to her own room and slept through the rest of the night. Judging from what happened, she must have been a very sound sleeper.

Mme. Achet slept at the back of the house, where there were three rooms overlooking the garden and accessible from it by means of a terrace. On the right was Mme. Achet's bedroom, in the middle was a little parlour, and on the left was the bedroom of Ali, the seven-year old boy. Two dogs usually spent the night under the terrace, but on this particular occasion they were absent. Mme. Achet accounted for this by explaining that "Sometimes they slept out." At this expression there was laughter in Court. So even the wretched dogs lost their characters in the case.

Up to this point, the narrative has been buttressed by the evidence of various witnesses, but now comes a span in which Mme. Achet was the sole authority.

After her maid left her, she went to her own bedroom and undressed. She then carried her clothes into the next room, which opened upon the terrace. There came a knocking on the window and she opened it.

> Q. "How came you to open it, when it was so late?"—A. "Someone called my name." Q. "Did you recognise the voice?"—A. "No, sir. I did not suppose it was M. Lépine who was knocking; I thought it was someone else." Q. "Who?" (No reply.) "You thought it was M. Thaunié?"—A. "Perhaps."

She opened the window and found herself face to face with Lépine, who wanted to come in. She set herself against this. (At this point in her evidence Mme. Achet burst into tears.) He declared that he wished to settle things up; but Mme. Achet objected: "Not at this time of night." None the less, Lépine persisted, and pulled out his revolver.

> Q. Was anything else suggested? Did he say anything to suggest that he had other grounds for coming?"
> Mme. Achet gave an affirmative nod.

She went to fetch her own revolver and got back to the window just as Lépine was clambering over the *barre d'appui*. She tried to push him back, whereupon he caught her by the arm, threw her off her balance, and somehow she fell on the terrace. She fired two shots with her revolver, whereupon Lépine let her go and ran off into the garden. She followed him closely, still firing; and at the last shot he tumbled down a declivity, falling about ten feet. She was in such a state that she "continued to run like a madwoman through the grounds," and, involuntarily, she came back to where Lépine was lying. She felt the body, and found that the heart had ceased to beat.

At this point it should be mentioned that when examined later by Dr. Sahut, Mme. Achet bore contusions on her forearm, on her arms, on her thighs, and on her right calf, contusions which the doctor considered as having possibly been received during a struggle such as she described. Strangely enough, Mme. Achet was much averse to this medical examination and submitted to it only after long persuasion, though the results corroborated her tale.

Before taking Mme. Achet's narrative any further, however, it seems desirable to draw attention to one or two points. In the first place, as the prosecution pointed out, if her object was merely to beat off an assault, why did she continue to pursue Lépine through the garden, firing as she went? On consideration, however, this is hardly so crucial a matter as it seems at the first glance. A woman in such circumstances might well lose her head and continue to discharge her revolver until it was empty. Or she might be so furious at the attack on her person that she might fire out of sheer rage, long after her assailant had taken to his heels.

A more serious point was raised when it was found that Lépine, in his flight, had blundered into an enclosed part of the garden instead of making straight for the side-gate. To account for this, the prosecution suggested that there was an accomplice on the scene who barred Lépine's direct line of escape, so that he diverged into the blind alley though he must have been familiar enough with the lie of the grounds.

More suspicious still was the admitted fact that Mme. Achet, from start to finish, raised no outcry. Questioned on this, she said that though her son was sleeping in the next room, he was a mere child and could be of no help even if she had cried out. This is true enough, and one could understand that a young mother would have no wish to call a seven-year old boy to witness the spectacle of herself at grips with an assailant. But the maid Françoise was on the premises and might have been roused. Mme. Achet's comment was that the maid slept a long way off and would have heard nothing. This sounds absurd, for Mme. Achet's residence was far from being a baronial castle; but she had some reason on her side, since it is a fact that Françoise Mosnier actually did sleep placidly through all the shooting which took place, and if she could do that, it seems doubtful if a scream would have awakened her.

There is, however, another possible explanation for the absence of any outcry from Mme. Achet: she may have had an accomplice whose presence would have been discovered if the maid had come down. But the whole question of a possible accomplice may be deferred for the present.

Meanwhile, it will be convenient to continue Mme. Achet's own account of what happened. Terrified by what she had done, she hit upon "the mad idea" of dragging the body out on to the road, so as to suggest that Lépine had never been inside the grounds of the house at all, but had met his death outside, at the hands of some night-prowler. It was raining in torrents. She went back to the house, secured the key of the iron gate, caught up a mantle to cover herself, and returned to Lépine, whom she dragged a distance of some fifteen yards and put on to the road outside the iron gate. Once she got him there, she took a knife which she used for gathering roses and with this she cut his throat. She then removed all papers from Lépine's pockets, and also his watch. With these she returned to her house, where she threw the watch and papers down the closet, cleaned the knife, extracted the empty cartridge-cases from her revolver, closed the window, and fell over-wrought upon her bed.

This portion of her story roused a certain scepticism. If the body were to be shifted at all, why not have removed it to somewhere near the terrace, in which case its situation would have tended to support the rest of her story? Again, there seems at first to have been some doubt as to whether the body had been carried or dragged along the ground. Lépine was a man weighing fifteen stones, a weight quite beyond the carrying power of a woman of Mme. Achet's physique; so that if any carrying had been done, the presence of a male accomplice would have been almost certainly established. In the end, however, the conclusion was in favour of the body having been dragged along the ground; and when tests were made with women of approximately the same build as Mme. Achet, it was shown that she might have dragged Lépine's body just as she described.

As for throat-cutting, the prosecution bluntly alleged that when she found Lépine he was alive, but dying from the effect of the revolver-shots; and that she cut his throat to stop the rattle, which might have drawn some passer-by to the spot, and not, as she declared, to suggest violence from some nighthawk. There was a considerable conflict among the medical experts as to whether a woman had strength enough to cut the throat of a goitrous man down to the larynx, and almost sever head from body; but it was finally agreed that a woman might have done it.

The fact that Mme. Achet removed the empty cartridge-cases from her revolver and hid them certainly supports her tale that she wished to throw suspicion upon some hypothetical *apache* who was to be represented as falling upon the notary and killing him with a knife for the sake of the contents of his pockets. But it was extraordinary indeed that she should have supposed that even the most superficial examination would overlook five bullet-wounds in the body, one of which had damaged the victim's eye. One must bear in mind that during this part of the affair Mme. Achet was, as she declared, "beside herself," and hardly likely to be capable of clear thinking. Further, the confusion of her doings at this stage certainly throws some doubt upon the contention of the prosecution that the whole tragedy was the result of a carefully thought out plan for drawing Lépine into an ambuscade of which every detail had been prepared beforehand.

Now comes evidence furnished by neighbours as to the events of the night.

Some doubt was thrown on the rather surprising somnolence of Mme. Achet's servant, who slept through the entire affair; but in Court she resented this violently and offered to take any oath that she was speaking the truth about it.

Martin, a carpenter, saw Lépine leave his house. About ten minutes later, Martin heard four shots and a loud cry from a man. Judging that it was no affair of his, the carpenter shut his window and went to bed. Three other neighbours seem to have acted in the same fashion.

More candour was shown by an old peasant-woman, the Widow Grandjean, whose house stood nearest to that of Mme. Achet. She went to bed about ten o'clock, after telling her beads, and was wakened by the noise of three revolver-shots and a loud cry, followed by the sound of a body falling.

> Q. "And you didn't get up?"—A. "Oh, no, sir! I pulled the bed-clothes over me; I was too frightened."

The most sensational evidence, however, was given by another neighbour, Melin by name.

> "On the night of the murder I got up about midnight to go to my dunghill, which is twenty-five or thirty

yards from Mme. Achet's house. Then I heard sounds of voices and stamping, as if someone had a man by the collar. This did not last long. The man was tripped up almost at once. He seemed to be held very firmly and threatened. I thought I heard: 'For mercy's sake! Leave off!' but I did not recognise M. Lépine's voice."

Questioned as to whether the threats were made by a man, Melin said that it was a man's voice. He heard sounds of heavy breathing, and thought that the man who had been tripped up was being dragged along the ground. Then, very distinctly, he heard a man's voice say: "Let's leave it there!"

Q. "You heard nothing further after the words: Let's leave it there?"—A. "No. I wanted to get back into the house again. It was raining, and I'd come out without my trousers."

All these various sounds, it appears, came from the roadway whither, as Melin learned next morning, Lépine's body had been taken. One must admire the skill which Melin showed in interpreting the noises he heard; but he was slightly off the mark, as later evidence seemed to show.

It was then the season for grape-gathering. A cultivator, Bataillet, was put into the witness box to explain that at the time mentioned by Melin, he and some friends were rolling a loaded truck to the press-house. Bataillet recalled distinctly that when they reached the press-house, one of the party said: "That's enough! Let's leave it there!"

Thus vanished what seemed at first to be one of the most convincing pieces of evidence produced to prove the presence of a male accomplice on the scene.

Meanwhile, all through the night, Mme. Lépine was anxiously awaiting the return of her husband, rising from time to time to go into his room, where the lamp still burned. One may ask why she did not take more active measures as the small hours drew on; but according to the prosecution she was afraid of a scandal and her fear was that Mme. Achet had detained Lépine with the intention of making him leave her house in broad daylight and so cause gossip. This does not suggest that Mme. Lépine had complete trust in the fidelity of her husband.

Be that as it may, in the dawn, towards 5 a.m., her fears got the upper hand. She rose from her bed and went to seek the cure, whilst her son Emmanuel set out in search of his father. Very soon afterwards he came back in high agitation crying: "They've murdered him!" Whereupon Mme. Lépine, half beside herself, ran to the crowd which had formed around the body, outside Mme. Achet's house.

In the meantime, Françoise Mosnier had shaken off her profound slumber, and, when she arose, she learned of the discovery of Lépine's corpse at the gate. She hurried to Mme. Achet's room, where she found her mistress still in bed. On hearing the news, Mme. Achet pretended complete surprise. "Good God, Françoise! What's this you're telling me?" She then seems to have got up and gone out on to the terrace which overlooked the crowd. Mme. Lépine, looking up, caught sight of her through her tears and cried: "M. Lépine was murdered in your house; you'll have to account for that!" Mme. Lépine heard no reply from Mme. Achet, who turned and went back into the house. Apparently she spent her time thereafter in burning some letters from Lépine, and also the acknowledgement of the 10,000 franc debt.

She despatched her little son, Ali, to her sister, and when Mme. Lamotte arrived, Mme. Achet asked her to take charge of the fatal revolver. "Do this for me," she said. "I'm under suspicion." Very unwillingly, Mme. Lamotte took the weapon, which later on she handed over to the police.

But this did not complete the list of Mme. Achet's little commissions. As it happened, her godmother—a Mme. Desgranges—was staying in Chantelle just then, though she had arranged to return home to Paris that night. Naturally, she betook herself to the house of her god-daughter. Mme. Achet received her with emotion and begged her to do her a favour. She handed a box of cartridges to her godmother, "in case things go wrong." She also asked her to go into her bedroom, find some papers which were there, and take them away with her. These, apparently, were the "certain papers" for the return of which Mme. Achet had been pestering Lépine at an earlier stage. What their contents were, no one can tell; for Mme. Desgranges burned them unread as soon as she got home. Finally, added Mme. Achet, "I have a sum of money which I want to give you. Look after it; you can give me it back later on. There are 6,000 francs hidden in the lining of the carpet, beside the chimney-piece in the parlour."

Mme. Desgranges obediently hunted out this money—6,100 francs in banknotes—and took it home with her. Next day, Mme. Achet was arrested, whereupon Mme. Desgranges—"fearing to be compromised"—took the notes back and stuffed them into their old hiding-place in the carpet, where, it is satisfactory to learn, they were in the end discovered by the police after all their wanderings.

All these quaint transactions occurred under the nose of a gendarme who was on duty in the house!

But the whole preliminary investigation by the police seems to have been astonishingly incompetent. Although Mme. Achet came under suspicion almost from the moment that the body was discovered, no steps were taken to secure her papers, no examination was made of her garments, though some of them must have been blood-stained, and she was allowed to communicate freely with her relatives, with the results described above. By the time the *juge d'instruction* arrived on the scene, the crowd had trampled out all trace of footmarks around the body, so that one possible means of settling the accomplice problem was completely destroyed. One of the gendarmes discovered a blood-stained cloth with which Mme. Achet had cleaned her shoes after the tragedy; he was told by his superiors that it was of no importance.

Political animus was imported into the case. M. Lépine had been the head of the Conservative Party in Chantelle. The chief of the Republican Party was the Mayor of Chantelle, Dr. Noir. On the sand of the garden paths were found numerous female foot-prints, but only one clear male imprint, a heel-mark, which may have been made by Lépine. Called in as an expert on account of his medical qualifications, Dr. Noir effaced this important trace—probably inadvertently, but certainly carelessly—before it had been properly examined. His political opponents at once accused him of having obliterated it intentionally, and a wrangle between supporters of the two opposing Parties seems to have ensued.

Another incident in the investigation produced a bad impression. Someone, described as an *agent de police volontaire*, was paid by the Lépine family to take little Ali Achet away from the house of his aunt, Mme. Lamotte, and bring him to the house of Mme. Lépine. Here the seven-year old child was questioned and his evidence taken down. It is not difficult to form an opinion on the value of a statement obtained in such circumstances.

Mme. Achet was arrested and imprisoned at Gannat. While awaiting trial, she wrote to her sister, Mme. Lamotte:—

> "Ah, well! This is the end, my dear Adeline, I can stand no more; I am suffering cruelly and I would rather die. I beg you not to make my poor little Ali responsible for what has happened. Be a mother to him, as I would have been to your daughter if this calamity had fallen upon you. Pardon me for my desertion; I can bear no more. . . . Farewell! By to-morrow all will be over."

Mme. Achet endeavoured to carry into effect the suicide which is hinted at in the above. Each day, during the winter, she set aside a piece of charcoal, furnished to warm her cell, and she hid the fuel in her mattress. When she had thus accumulated a store which she reckoned as sufficient, she rose one night, removed the chimney of her stove, stopped up the chinks in her cell, and lay down again to await asphyxiation. She was, however, discovered in time and revived.

At this point, it seems advisable to consider some problems which even the trial of Mme. Achet left unsolved.

Plainly there were two keys to the mystery, and both of them are missing. One may well believe Mme. Achet's statement with regard to her interview with Lépine on October 5: "If I were to tell you the circumstances, you would know the secret of a good many things." We should know, for instance, whether Lépine behaved purely as an honest notary concerned only with business affairs, or whether he had ulterior designs of one sort or another; and if we had that knowledge, we should be very close to the centre of the problem.

The second key, obviously, lay in those "necessary documents" which Mme. Achet demanded from Lépine. He brought them with him on the night of the tragedy, for no trace of any such papers was found among his effects after his death. Mme. Achet was careful to remove them from his pockets after she had shot him; but once she had them in her possession, she was eager to rid herself of them; and they were finally burned, unread, by her godmother, Mme. Desgranges. Unless these documents dealt with really vital matters, it is hardly credible that so much importance should have been attached to them by the only living person who knew what they contained.

These two keys being lost, one is reduced to mere speculation in seeking for the motive which underlay the facts already given above. The simplest suggestion is that Mme. Achet, having been driven into financial difficulties, murdered Lépine in order to extricate herself. Can one reasonably accept this as satisfactory? Mme. Achet had one rich and generous lover, Thaunié, and there were others. as well at her beck and call. With such resources at her back, is it likely that she murdered Lépine for the sake of clearing herself of a debt of 17,200 francs? It would surely have been simpler to borrow the sum elsewhere and so tide herself, temporarily at any rate, over her difficulties. Besides, she had no means of knowing what entries Lépine made in his account books at home. For all she could tell, he might have left a complete statement of her financial liabilities behind him, a statement which could not be obliterated by murder. Actually, as has been seen, he did leave an entry dealing with the 7,200 francs which he "forgot" in her house on October 5.

If the financial motive be discarded, something else must be found. Mme. Achet's own suggestion was, of course, that Lépine was enamoured of her and was determined to possess her by fair means or foul. Against this may be urged the evidence of Lépine's own wife and son, who testified that he spent the evening before the tragedy in playing cards with them and that he was perfectly calm at that stage. Would he have been as cool as all that if he intended to leave the card-table in order to play Tarquin to Mme. Achet's Lucrece? On the other side one must not overlook the possibility that Lépine was a good actor, or a man with strong self-control, or that his wife and son were not particularly observant.

There is, however, yet another possible motive which deserves consideration. Strangely enough, it never seems to have crossed the minds of either the prosecutor or the counsel for the defence, for although it can be condensed into a single word, neither side mentioned it in the trial. That word is "Blackmail."

Let us consider this a little more closely. Here are two parties in the case: a man and a woman; and it is possible that the woman might blackmail the man or *vice versa*. Is it likely that Mme. Achet was blackmailing Lépine and that she extorted from him 7,200 francs on October 5 by putting some screw on him from her private knowledge? A moment's consideration of the evidence is enough to disprove any such hypothesis. It was Lépine who was pressing for a settlement, a final settlement; it was Mme. Achet who was "dancing on a spring-board," not knowing where she would land. It was Mme. Achet who was insistent on securing those mysterious documents which finally vanished, unread. These papers may have contained something which would have put out of Mme. Achet's reach, once and for all, any possibility of securing Thaunié as a second husband. This is the purest speculation, but one may be permitted the surmise.

Then there is the question of premeditation. Was Lépine drawn into a pre-arranged trap or not? The prosecution pointed to the haste which Mme. Achet displayed in getting her revolver from Paris, and they insisted that this proved the existence of a preconcerted scheme of murder. But did it? Mme. Achet might equally well have been in fear of physical violence from Lépine and may have wanted her weapon for self-defence.

More to the point is the manner in which Lépine entered the premises. Had he come on a purely business errand, his natural course would be to ring the bell at the iron gate in front of the house and wait to be admitted either by the maid or by Mme. Achet. This, apparently, he did not do. If, instead, he climbed over the railings it was a curious way of coming to pay a business call. Remains the little side-gate. Various neighbours testified that as a general rule, this side-entrance was locked at night; but it was found open on the morning after the tragedy. Why was it opened? For the convenience of Lépine? Or did Mme. Achet leave it unlocked for the use of some lover whom she was expecting? According to Mme. Achet's own evidence, when she heard somebody at the window calling her name she did not think of Lépine. "I thought it was someone else," she declared.

This brings up the problem of an accomplice; and it is interesting to note that the prosecution, up to the very last moment, adhered to the hypothesis that one or more accomplices were on the spot at the critical moment. But the prosecution had swallowed Melin's evidence holus-bolus, and they even quoted it against Mme. Achet in the final speech
made on their behalf, despite the plain fact that it was completely shattered by the testimony of Bataillet. From this it may be inferred that the presence of an accomplice was a fixed idea in their minds, independent entirely of the statements of witnesses.

Apart from Melin, they relied on the testimony of seven-year old Ali Achet, whose evidence was taken on commission. It was to the effect that he had heard Lépine cry "Holà! Holà!", then some whisperings, the voice of an unknown man, and noises of different people walking. Apparently the noise of the revolver-shots had not struck him. One may assess this evidence at what it is worth, bearing in mind the incident in which he was taken to Mme. Lépine's house and questioned there.

In considering the possibility of an accomplice, one must bear in mind the undenied fact that during the struggle at the window,

Mme. Achet "fought mute" when one might have expected her to cry for help; and this despite the certainty that the reports of her revolver were as likely to attract attention as any screams.

No one can pretend that Mme. Achet was a wholly reliable witness on the affairs of that night. In her initial statement, during the first interrogations, she told a story which she subsequently withdrew completely. According to this tale, she fainted after the struggle on the terrace. When she revived, she saw standing by her bed a man engaged in perusing some blood-stained papers. This man had defended her against Lépine, had struck the blow with the knife, and had helped Mme. Achet to carry the body into the road. The general view seems to have been that this story was the purest fantasy.

Who was the accomplice, if accomplice there were? Various possibilities present themselves, and they are worth examining, even if they are absurd.

The accomplice might have been a confederate of Mme. Achet in an attempt to blackmail Lépine. On this assumption, she must have invited him to her house to provide a witness whilst she got Lépine into a compromising situation. But there is not the slightest proof that she ever intended to blackmail the notary.

Again, the accomplice might have been a lover visiting Mme. Achet by night, who was surprised when Lépine arrived and detected his presence. On this basis, jealousy would lie at the root of the tragedy. But Mme. Achet bore scratches and bruises which seem to have been the result of a struggle. Is it likely that any lover would stand by whilst his mistress was thrown down on the terrace before his eyes?

Yet again, the accomplice might have been introduced purposely by Mme. Achet as a defender, if she feared a nocturnal attack from Lépine. But any lover would have shrunk from actual murder in the circumstances, and he would certainly have been able to inform Mme. Achet that the elimination of Lépine himself was no way out of her financial troubles, unless the notary's books were also destroyed.

Finally, the accomplice might have been some *apache* or other hired by Mme. Achet. But a woman of her class was unlikely to have relations with the underworld; one cannot go out into the street and hire the first ruffian who comes along.

With that, one may leave the problem of this hypothetical accomplice.

For three whole days, the jury listened to the evidence. A fourth day was devoted to the speeches of the prosecution and the defence. At the end of this lengthy business, the jury deliberated for a quarter of an hour and returned to court with their verdict.

They found Mme. Achet guilty, but with extenuating circumstances, and they dismissed the suggestion of premeditation.

The Judge sentenced Mme. Achet to twelve years' penal servitude. He awarded the Lépine family 2,000 francs (£80) instead of the 17,000 francs (£680) which they claimed.

Thus, the jury rejected any idea of a prearranged ambuscade, whilst the Judge, by awarding only 2,000 francs damages, made it clear that he was far from satisfied about the late Lépine's manner of conducting business with Mme. Achet.

But the tragedy was not completed when sentence was delivered in the court at Moulins on May 3, 1891. In January of the following year, Mme. Achet showed symptoms of madness, and had to be placed in a lunatic asylum.

One may well ask when Mme. Achet became unhinged. She showed no sign of derangement during her married life, nor during the subsequent period when she entertained her numerous lovers. Were the seeds of her madness sown in that nightmare experience at the window and the wild chase through the rainy garden? Were they ripened by the strain of her imprisonment and trial, until they brought out their final fruit in gaol and in the madhouse? None can tell. An impenetrable mystery hangs, and will hang for ever, over all this tenebrous affair.

II
WHERE PLOTS COME FROM

WHERE DO THESE AUTHORS get the ideas which they develop into the tales we read? A puzzling question, this, though not beyond all conjecture in some cases.

To begin on the lowest level, there is the poor gentleman, smitten with the itch to write, but with no ideas whatever, and no chance, apparently, of extracting any from his internal resources. One might suppose his case to be hopeless. But far from it. Someone has already foreseen his difficulty, forestalled his needs, and produced for him a handbook containing all the plots which he could use if he wrote steadily from year's end to year's end. Indeed, if my calculations be correct, there would still be sufficient material left over to let him start afresh on New Year's Day. How is it done?

In my early childhood there was in vogue a game for tiny tots which served its purpose for a time but is now long forgotten. Three packs of cards were employed in it. On the faces of the first pack was printed an assortment of heads; the second pack bore pictures of bodies; whilst on the third were drawings of legs and feet. A card from each pack was dealt to the player; and by fitting together the three fragments which fell to his lot, he might obtain as a reward either the portrait of a young Adonis or, more probably, the depictment of a creature with a snake's head, the body of an ostrich, and the forelegs of a hippopotamus. The latter result raised screams of laughter, for we were simple little souls in those days, and easily amused.

So much for the game of "Heads, Bodies, and Legs" of happy memory. Turn now to Aristotle, who believed—perhaps not incorrectly—that a story should have a Beginning, a Middle, and an End. The principle on which the *Idea-less Author's Guide* is compiled should now be obvious; but it may be well to give a concrete example of how the work is used.

Like Gaul, the volume is divided into three parts, containing respectively what may be termed Heads, Bodies, and Legs. As in works on algebra, the factors are entitled A, B, C, and C^1, where A stands for the heroine, B for the hero, C and C^1 for third parties according to sex. The eager seeker for plots looks through the first section and chooses something like this:—

HEAD
A, rich but ugly, is courted by B, poor but handsome.

At the end of this, the inquirer notes certain figures which refer to the middle section; and, following up the clues, he finds himself faced with this:—

BODIES
(*a*) They marry, but B wearies of A.

(*b*) They marry. B makes a great invention which renders him independent of his wife's fortune.

(*c*) They marry. A gets her face lifted and grows beautiful.

(*d*) They marry. B is attracted by C and neglects his wife.

After each of these are further figures guiding the reader to the final section. Suppose, for instance, that the would-be writer selects BODY (c), then he will be led to:—

LEGS
(1) They live happily ever after.

(2) A is attracted by C^1 and B by C. They drift apart and end with divorce.

(3) B is disfigured by lightning, whilst A loses her money. Reduced to even terms, they make the best of it.

But enough has now been said to show that no one need shrink from authorship merely because he has not enough originality to invent a plot for himself.

Another method of plot-construction lies ready to the hand of anyone; and it is probably the commonest one chosen by the beginner-novelist: the autobiographical method in which the hero is simply the author himself under a thin disguise, and his actual life-history furnishes the backbone of the narrative. Dickens did not despise this system, as we know from *David Copperfield*. The drawback is that most people's lives are so uneventful that they would interest only themselves; and a book with a circulation of a single copy is not likely to be worth printing.

There is, however, a simple variation, even more pleasing to the egotistical writer. The author is his own hero, as in the last example; but instead of describing his real life, he depicts the life he thinks he ought to have led if only Fate had given him a square deal and treated him according to the merits which he feels he possesses. Budding authors who adopt this system run the risk of earning the criticism passed by a candid friend on a flattering photograph: "It's not like you. But it's like what you would like to think you look like." Samuel Butler liberated his spleen by combining both methods in *The Tray of All Flesh*, in which the history of Ernest is practically pure autobiography, whilst Overton is an idealised portrait of Butler—Butler as he thought he might have been if he had been given fair treatment.

Rather wider scope is given by the method of selecting a general theme in the first place and letting the plot arise from that, whilst in turn the characters are suggested by the plot. Charles Reade is probably the most successful exponent of this system in English. Greatly moved by the evils of the lunatic asylums in his day, he fixed on them as his subject. The immense scrap-books of newspaper cuttings which he accumulated were drawn upon for facts, and out of them grew his plot, with the characters necessary to illustrate it. So he constructed *Hard Cash*. In *It Is Never Too Late To Mend* he complicated the matter further by combining two main topics: the defects of the prison-system and the gold-rush to Australia.

But in studying this "documentary method of plot-fabrication, the best exemplar is Zola, for it chances that we have a very full description of how he went to work. Take, as an example, his twenty-volume history of the Rougon-Macquart family. In starting a fresh tome of that colossal enterprise, his first care was to choose the background for the tale he was going to tell. It might be a big shop, or the Halles of Paris, or a coalmine, or a railway.* It might even, as in *Pot-Bouille*, be a set of flats occupied by bourgeois families. ("A strange world," as one critic acidly remarked, "where concierges speak like poets and everyone else talks like a concierge.")

The next step in Zola's procedure was to accumulate the fullest possible information about the environment which he had selected. He read books; he consulted experts; if he was going to write about a railway he made a trip on the foot-plate of a locomotive; if a coalmine was his subject he went down the shaft and took notes on the work of the men at the face. All the material thus acquired was then classified and catalogued, forming what he termed the *dossier* of the book.

This done, Zola sat down and wrote what he called an *ébauche*, which was really a long, chatty letter to himself, for he thought best with a pen in his hand. It was during the writing of this *ébauche* that his plot began to take shape in his mind; episodes suggested themselves; characters came into being to fit themselves into the shadowy outlines of the story.

Then came yet another stage in which Zola drew up a set of documents dealing with the physical and mental characteristics of

---

* Curiously enough, a very different type of author chose the same system: R. M. Ballantyne, whose books for boys delighted a whole generation.

each character in the book and the part which it played in the plot. This voluminous collection was labeled Personages.

Only after all this toil had been completed did Zola turn to the actual allocation of his material into chapters. Episodes were drawn from the pages of the *ébauche;* the characteristics of the actors came from the personages file; technical information was extracted from the *dossier:* and the whole material was put together and divided up as required. Thus, at long last, Zola had his plot complete in every detail.

But even at this stage there came a complication. Zola was a creature of crotchets and whims; and one of his fads was that in a book each chapter should contain roughly the same number of pages. Hence it was sometimes necessary to condense one section of the narrative or expand another so that both should fill the same space in print. One marvels at the patience and the labour involved in such a method. How any man had the tenacity to carry it through the twenty volumes of the Rougon-Macquart series, the three books *Lourdes*, *Rome*, and *Paris*, and the set of *Four Gospels*, is a source of continual astonishment.

In the Zola method, the plot comes first in a shadowy form and out of it arise the characters which it demands. Possibly as a result of their evolution, these characters tend to be typical rather than individual. What about inverting the process? Start from characters and develop a plot which has its roots in their idiosyncrasies?

Mr. Somerset Maugham, in those prefaces which form not the least interesting parts of his writings, has let us into some of the secrets of his workshop. Take the case of *Rain*. Traveling from Honolulu to Pago-Pago, Mr. Maugham saw among his fellow-passengers three people: a missionary and his wife and a prostitute supposed to be flying from trouble in Honolulu. "I talked with the missionary and his wife but once," he says, "and with Miss Thompson not at all." But in his methodical way, he jotted down notes of the appearance of each of them; and in his single conversation with the wife, he learned a good deal about the past history of herself and her husband, a singularly unlikeable couple if one may judge from his notes.

Later, evidently, he began to reflect upon the possible reactions which would occur if this trio got cooped up together for any length of time. The missionary, being what he is, would persecute the prostitute; and she, being what she is, would fall temporarily into religiosity. Then would come tragedy, arising directly out of the characters in the tale. Stated thus, anybody might have written the story; one needs Mr. Maugham's tolerant cynicism and dry, incisive style to convert that bare skeleton into the little masterpiece: *Rain*.

In his preface to the trilogy: *Whom God Hath Sundered*, Mr. Oliver Onions describes the genesis of his novel, and it makes a very curious and instructive story. Initially, he sat down to write a tale of about 5,000 words in length for a magazine. Looking up his notebook for a basic idea, he came upon this: "Under guise of speed exercise and subsequent transcription young student of shorthand gets confession of suicide from fellow-student—" etc. So Mr. Onions began his short story of a murder. But Mr. Onions is one of those fortunate writers whose characters seem to "come alive" at once; and one surmises that, having "come alive" in this case, they demanded more space than could be given to them in 5,000 words. As a result, he had to abandon the short story and expand the material into the elbow-room provided by the frame of a novel. This he did, and the result was that extremely clever tale of a murder which appeared under the title: *In Accordance with the Evidence*. This gives the story of the crime as seen through the eyes of the murderer, and it ends with his reflection on the murder. "They say somebody always pays. . . . No: nobody has paid. Nobody ever will."

But even while he was writing *In Accordance with the Evidence*, Mr. Onions found, shaping itself in his mind, the question: Can a man who commits a murder remain precisely as if he had not committed one, and, if not, in what respects is he different? No sooner had he finished *In Accordance with the Evidence* than he found himself beginning *The Debit Account*, in which he answered his own question. A murderer can never be the same as an innocent man, for his every word and every action must be governed by the necessity of averting suspicion from his past. Even in his most casual conversation he must be on the alert to see that he drops no clue which might lead someone back to the crime. It is this state of perpetual strain and apprehension, as recorded by the murderer, which forms the basis of *The Debit Account*. And, when the balance has been struck, it turns out that the murderer's boast took no account of the complexities of his position. Not only he, but other people as well, have eventually to foot the bill.

No sooner was *The Debit Account* well under way, than Mr. Onions found himself faced with his old problem in a new guise. Hitherto the events and characters had been seen through the eyes of the murderer. He had committed his crime in order to shield the girl with whom he was in love and to secure her for himself; and naturally to him she appeared all-desirable. But was she, in reality, so wonderful after all? As Mr. Onions says in his preface: "Both these stories were related by the chief actor in them, Jeffries himself, and were obviously partial and incomplete without some independent comment from outside. The only other character capable of furnishing any comment of value was the one of whom least of all had been said—Louie Causton."

So, in order to round off the narrative, Mr. Onions had to take up his pen again and write the third part of the trilogy, *The Stag of Louie*. By couching it in the third person, he was able to treat all the characters objectively and assess each of them at a proper valuation. And in this final section of the story there is the complete balance sheet of the transaction in murder.

Turning again to his preface, we find Mr. Onions making a plain confession. "From the foregoing it would appear that only from time to time was I myself enlightened as to the direction in which I was moving—or rather, perhaps, as to what my next step was to be. At the time I regretted this. I should have liked to say of the whole that it had been deeply plotted and designed before pen was put to paper." Few people would have guessed this from an examination of the final form of the book *Whom God Hath Sundered*, in which sections of the original three independent volumes are interleaved to yield a connected narrative. The only alteration demanded was the deletion of "one repetition in the Envoi" (necessary at the time). Apart from that, the whole text fits together with the accuracy of a jig-saw puzzle, and no one would guess that when he began it Mr. Onions had not every incident already in his mind. As an example of how a plot may be developed from a few initial characters, the three volumes and the resulting one-volume trilogy make a most fascinating study for anyone interested in this aspect of literature.

Yet another form of plot-genesis can be traced, wherein the initial inspiration comes from events and not from characters. In the pages of that long-extinct magazine, *The Idler*, Sir Arthur Quiller-Couch described how he picked up, piece by piece, the facts which went to the making of *"Dead Man's Rock,"* and how he invented the thread which links them together in the structure of the tale.

To take the title first, it seems that during a seaside picnic on the Cornish coast, "Q's" attention was called to a pillar-like rock standing up from the sands and veined with curious red streaks, like bloodstains. "I want a story written about that rock," said one of the party. "Something really blood-thirsty. 'Slaughter Rock' might do for the name." But "Q" chose better when he took his title from the Dodman, a promontory east of Falmouth; and he did still better in taking the local pronunciation, Deadman, which brings with it the hint of the macabre.

Sir Arthur's father wrote a history of Polperro, in which he mentioned a curious incident in the Quiller family history.

"In the old home of the Quillers, at Polperro, there was hanging on a beam a key, which we as children regarded with respect and awe, and never dared to touch, for Richard Quiller had put the key of his quadrant on a nail, with strong injunctions that no one should take it off until his return (which never happened), and there, I believe, it still hangs." Musing over this anecdote of his great-grandfather, Sir Arthur saw that, for literary purposes, this key should be made to unlock something much more valuable than a sea-captain's quadrant; and he evolved the germ of a plot in this form:

"A family living in poverty, though heirs to great wealth—this wealth buried close to their door—and the key to unlock it hanging over their heads from morning to night."

But what was the treasure to be? As luck would have it, Sir Arthur had a taste for reading old books on travel: Maundevile, Marco Polo, Hakluyt, Purchas, etc., and during his reading he came upon the Great Ruby of Ceylon. Add a curse to the jewel, and the interest rises.

Now a single-handed treasure-hunt is weak stuff. Competition and conflict are needed, especially if the curse is to be given a sporting chance. So "Q" added a further section to his plot. The treasure was to be buried by a man who had murdered his comrade and sole confidant so that he might enjoy the treasure alone, and had afterwards learned of the curse attached to the booty. The descendants of the victim and his murderer were to be rivals in the search for the treasure, each side possessing half of the essential clue. How was this last condition to be plausibly fulfilled? Sir Arthur invented a buckle—a golden buckle, of course, since it looks better and costs only the trouble of inserting an adjective in the manuscript—and this buckle had the clue engraved straight across the two halves when joined together. Thus each part by itself was meaningless. Place a half in the hands of each competitor and the thing is done.

Even at this stage, however, "Q" was unable to find an opening for this book; and it was the purest chance which threw into his hands the facts which he needed to give him a satisfactory start. He happened to spend some weeks at the Lizard, and he took with him some guidebooks and local histories with the idea of learning something about the locality. Amongst these was a *History and Description of the Parish of Mullyon*, by its vicar; and it was in this work that "Q" unearthed the historical incident which formed the basis for the early chapters of "*Dead Man's Rock.*" They were strange enough in themselves, without any adornment from fiction.

On March 25th, 1867, a Dutch barque was wrecked on the coast and every soul aboard her was drowned, with the exception of a Greek sailor, who was found clambering on the rocks next morning.

An inquest was held, and this Greek was one of the witnesses. On being questioned about the lost vessel, he answered: "I joined the ship at Batavia, *but I do not know the name of the ship or the name of the captain.*" When shown an official list of Dutch East Indiamen, he picked out the name of the *Kosmopoliet* (Captain Konig), saying that this was the lost ship. He then gave his evidence on the disaster, which—as he was the sole survivor—passed without contradiction; and thereafter he vanished from the district.

Almost immediately after his disappearance, the Dutch Consul at Falmouth (Mr. Broad) arrived, accompanied by the captains of two Dutch East Indiamen then lying in that port. One of the captains asked: "Was it Klaas Lammerts' ship?" On being told that the wrecked vessel was the *Kosmopoliet*, he was incredulous. "I don't believe it; the *Kosmopoliet* isn't due for almost a fortnight. It must have been Klaas Lammerts' ship." At this stage the Vicar himself appeared, with a scrap of flannel he had picked up on the beach, bearing the inscription, "6.K.L." "Ah!" said the Dutch captain, "It must have been the *Jonkheer*." Any further doubt on the point was removed by the discovery on the shore of a parchment which turned out to be the masonic diploma of Klaas Lammerts. Thus the sunken vessel was definitely identified as the *Jonkheer Meester van de Wall van Puttershoek*, Captain Klaas van Lammerts, 650 tons register, homeward bound from the East Indies with a cargo of sugar, coffee, spices and some Banca tin. The value of ship and cargo was about £40,000 to £50,000.

Obviously the Greek sailor had either lied or been mistaken over the name of the ship in which he had sailed from Batavia; and his sudden disappearance raised some suspicions. It was recalled that on the afternoon before the wreck the ship had appeared to manoeuvre in a very clumsy fashion, and had missed stays several times in her efforts to beat off the coast. An impression gained ground that there had been a mutiny on board, and no doubt the value of the cargo lent strength to suspicion. But the disaster remains one of the many mysteries of the sea.

If the reader recalls *"Dead Man's Rock,"* he will now see how much of the history of the *Belle Fortune* came direct from reality:

the wreck on the Cornish coast, the confusion of identity between two vessels, the hints of violence during the ship's last hours, the Greek survivor, and even Simon Colliver's astounding lie: "The ship belonged to Bristol, and was homeward bound, but I know neither her name nor that of her captain." Truly "Q" was fortunate in having such material thrust into his hand by the veriest accident.

Fame is a curious business. In a little East Coast town in Scotland, a monument commemorates an adventurer, son of a local cobbler, who, when summoned in his youth before the kirk-session of his day for indecent behaviour in church, "did not compear, being gone away to the seas." Had that young man not joined the *Cinque Ports*, or had he managed to avoid quarrelling with Captain Straddling, he would never have been put ashore at his own demand on Juan Fernandez; and if, after his rescue from that island by Captain Woods Rogers, his story had failed to reach Daniel Defoe, the world would lack its "*Robinson Crusoe*," and Alexander Selkirk would probably have no statue to honour him in the streets of Largo.

"*Dead Man's Rock*" may count as a novel of adventure, but it brings us to the borders of the novel of crime, which includes the detective story as one of its varieties. How can one set about devising a plot for a crime story?

Begin with a procedure expounded by Poe in a footnote to his "*Mystery of Marie Rogêt*"; though in dealing with such a deliberate mystifier as Poe, one hardly knows whether to take his statements seriously or not. Poe declares that under pretence of solving the mystery attaching to the fate of a Parisian *grisette*, he followed in minute detail the essential (while merely paralleling the inessential) facts which came to light about the real murder of Mary Rogers, no solution of which had been found at the period when Poe published his tale. "Thus all the argument founded upon the fiction is applicable to the truth; and the investigation of the truth was the object." Poe further claimed that confessions made by two persons long after the publication of his story "confirmed in full not only the general conclusion, but absolutely *all* the chief hypothetical details by which that conclusion was attained." Believe him or not, just as you please. But here, at any rate, is one method of finding a plot; and it could be applied to other cases where the mystery of the crime is still unsolved. Not everyone, however, can bend the bow of Ulysses.

Few writers would care to rival Poe in the extreme meticulousness with which he applied his method; but the principle may be utilised with less rigidity by taking the broad facts of a real case and transposing them into fiction with suitable modification. Mark the qualification. After all, there is such a thing as the law of libel. And fiction must present a semblance of probability from which real life is often completely disparted.

Look, for example, at that splenetic book, *L'Immortel*, in which Daudet poured ridicule on his aversion, the Institute. In the novel, the protagonist is the Perpetual Secretary of the French Academy, Astier-Réhu—"vir ineptissimus," as the illustrious Professor Dr. Schwanthaler of Bonn would say. He is gulled by a dissolute little bookbinder, who sells him a whole series of forgeries purporting to be autographic documents from historic celebrities. Doubt is cast on the authenticity of these things. In one of them the Emperor Charles V addressed Rabelais as "Maître Rabelais" instead of as "Frère Rabelais"—"as if Emperors never made slips of the pen," as Astier-Réhu comments bitterly, when the blunder is pointed out. Finally, the whole affair explodes in a trial of the bookbinder for forgery, a trial which covers the Perpetual Secretary and the whole Academy with ridicule. It is all kept within the bounds of probability.

Compare this with the facts on which Daudet based his novel. In real life, the prototype of the fictional Astier-Réhu was a M. Michel Chasles, an eminent mathematician of his times and a member of the Institute of France. The forger was one Vrain-Denis Lucas, a man of no great education who must have had the most complete contempt for the intelligence of his client—a contempt which proved itself to be thoroughly justified by results. Between 1861 and 1870, Lucas succeeded in foisting upon Michel Chasles no fewer than twenty-seven thousand of these forgeries, reaping some 150,000 francs (£6,000 in those days) for them, a sum which was practically all profit to the forger since his expenses were confined to the moderate cost of ink and paper. This works out at a shade over eight forgeries per diem purchased by the gullible M. Chasles over a period of nine years.

It is, however, when we examine the nature of these forgeries that we see how modest Daudet was in comparison with the reality. One or two samples will suffice: a letter from Lazarus to St. Peter, written on paper, in French; a note from Sappho, ditto, ditto; an epistle from Mary Magdalene to some King of the Burgundians, ditto, ditto. One can understand the forger's use of paper; for, working on his scale, the necessary parchment would have demanded the flaying of flocks of sheep. But how any educated man could have imagined that Cleopatra corresponded with Caesar in French is incomprehensible. Someone once spoke of "the extreme simplicity of a child or a mathematician," and, with Michel Chasles in mind, one may admit the aptness of the description. What is less intelligible by far is how Vrain-Denis Lucas escaped writer's cramp.

This method of taking an actual case as a basis has been practised in England with most success by Mrs. Belloc-Lowndes. Jack-the-Ripper's career furnished her with the foundation for her book "*The Lodger*," in which she attempted to account for Jack's safety by depicting him as a religious maniac sheltered by a compassionate landlady. In that other gruesome novel, "*The Chink in the Armour*," she chose as her subject the Goold case in Monte Carlo, and showed "how it might have happened."

On the whole, however, it takes a practised novelist to make a success in this field, in which divergence from the actual facts is sternly reduced to a minimum. Unless a skilful hand is at work, the result is too apt to give the impression of someone dancing in fetters. A much easier procedure is obtained by taking a case from real life, musing over it, turning it this way and that, until in the end one has evolved from the facts a plot which has practically nothing in common with the starting-material except some details surviving from the evidence in the original trial.

In my day, I have spent a good deal of time in studying innumerable criminal cases, and in the course of this I came across

M. Albert Bataille's admirable volumes which contain his own reports and impressions of French trials from 1880 onwards under the title: *Causes criminelles et mondaines*. They are not so accurate as our own series of 'Famous Trials' but they have one advantage over ours, for M. Bataille placed himself to some extent in the shoes of a juryman, so that his description of a trial gives one some inkling of how the case appeared day by day from the jury's standpoint, with the mental effect of the evidence, pro and con, set down for us to see.

Amongst M. Bataille's cases, I came across the Chantelle mystery, an account of which is given earlier in this book. As it stood, it was of no use to me; but it served as a starting-point; and I turned it over and over in my mind, elaborating here, excising there, until all that remained of the original material was swamped by pure inventions. I retained Mme. Achet's house and garden, but transferred them to this side of the Channel. I invented a new life-history for her, and gave her a companion and a pet as well. Also some people appeared out of the blue to play their parts in a fresh plot which was coming into existence in my mind. In the murder, I kept as closely to the original evidence as possible, since it is well to have a reliable basis for one's technical details. Then a second murder forced itself into the scheme and demanded something entirely fresh in the way of alibis. And so it went on, until very few people would recognise the actual root from which my "*No Past is Dead*" really grew. Anyone interested—if such a creature there be—can compare the novel with the details of the real-life tragedy as given in these pages.

Even Conan Doyle at times did not disdain to take some hints from the world about him. "*The Hound of the Baskervilles*" owed its inception to a remark by Fletcher Robinson that a spectral dog haunted the neighbourhood of his home on Dartmoor; whilst Part II of "*The Valley of Fear*" is based on the actual experiences of a young Pinkerton detective during the Molly McGuire outrages in the Pennsylvanian coalfields.

As for Sherlock Holmes himself, it is plain that he did not spring fully-armed from his creator's brain, like Minerva from the head of Jove. His thought-reading propensities—so bewildering to the illiterate Watson—come straight from a similar characteristic in Poe's Dupin, described in *The Murders in the Rue Morgue*. Indeed, one would like to hear Poe's views on the matter, for he was super-touchy on the subject of imitation. Then, again, Holmes owes more than a little to Conan Doyle's memories of his old teacher in Edinburgh, Joseph Bell, as Sir Arthur himself gratefully acknowledged. The inferences from the hat in *The Blue Carbuncle** and the conversation of the Holmes brothers as they sit in the club window, as given in *The Greek Interpreter*, come straight from the Bell mint. But hundreds of students must have listened to the diagnoses of Joseph Bell. It took a Conan Doyle to appreciate them and to apply them in a fresh field.

On looking back over the short stories on Sherlock Holmes, one cannot but be struck with the variety of the problems presented in them and by the paucity of the murder motive amongst them. On the other hand, every one of the four novels turns upon a murder. One may infer from this that as a general rule it is necessary to put the supreme stake—a human life—on the table if an author is to retain the interest of his reader through the long stretch of a full-length novel. And as the volume of short stories nowadays seems to have a smaller public than a long novel, one may further infer that the apparent increase in "bloody-mindedness" among the authors and readers of detective stories arises from the modern numerical preponderance of the long narrative rather than from any growth

---

* It is not generally recognised that Conan Doyle had in Beaumarchais a forerunner quite as ingenious as himself in this particular field and earlier by at least a century. In his *Gaieté faite à Londres le 1er Mai*, 1776, Beaumarchais drew a series of inferences concerning the owner of a lost opera-cloak which are as good as those made by Sherlock Holmes after inspecting Mr. Baker's hat. It is doubtful if Conan Doyle ever read Beaumarchais, so no suggestion of imitation is involved.

of brutality in the public mind. For the present, at any rate, an author with a plot depending on something less than murder would be well advised to handle his subject in short compass, whilst if he wishes to make murder his theme, he had better launch into a full-length effort.

But enough of the novel of crime. It is time to examine a fresh aspect of the subject. Another source of plots is fondly supposed—by the uninitiated, at any rate—to lie in Dreamland. There, it seems, nuggets a-plenty are believed to lie broadcast, costing but the trouble of picking them up. Some people even keep notebooks by their bedsides so that, when they awaken from some vivid adventure in ultimate dim Thule, they may immediately jot down a word or two which will remind them of it after the dawn. But very probably most of them fare no better than their fellow who, opening his night-inspired tablets in the light of day, read the pregnant sentence: "A strong smell of gas pervaded the kitchen."

Probably the most widely-quoted instance of a helpful dream is that which concerns R. L. Stevenson's *Strange Case of Dr. Jekyll and Mr. Hyde.* Usually the legend takes the form: "Stevenson dreamed the complete plot of the story and wrote it when he awoke." It is, perhaps, a pity to destroy so pretty a fable, especially as it has gained so wide a currency. But truth is truth, after all, and Stevenson himself should be the best authority on the subject. His own account is simply: "I dreamed the scene at the window, and a scene afterward split in two, in which Hyde, pursued for some crime, took the powder and underwent the change in the presence of his pursuers. All the rest was made awake and consciously." Readers of the book will realise how much was supplied by the waking imagination and how little by the dream.

III
STEPS TOWARDS THE ATOMIC BOMB AND BEYOND IT

IN MY STUDENT DAYS, at the end of last century, we had very definite ideas about atoms and elements. When we thought of an atom, we pictured in our minds something like a ball from the bearings of our bicycles, reduced to infinitesimal dimensions: a tiny sphere, harder and more resilient than steel, wholly incompressible, unchanged since the dawn of the Universe, and unalterable down all the vistas of the future. As for the elements, they were composed of congeries of these atoms; and in each particular element, all the atoms were identical with each other in weight, size, density, chemical properties, etc. Thus, in a piece of pure gold, every golden atom was precisely like every other atom in the material; in a block of pure iron, every iron atom was identical in every respect with all its neighbours. But a golden atom was something quite different in its nature from an iron atom, as could be seen from the different physical and chemical properties of the two metals. Using a crude analogy, one could liken a specimen of one element to a mass of swan-shot, whilst another element could be represented by a heap of marbles, and a third by a set of billiard-balls.

I was a modest young animal. Little did I think in those days, as I absorbed the lore of my teachers, that I should myself have a share in upsetting these notions, which then seemed so firmly grounded.

As the mariners of an earlier age relied in their voyagings upon compass, log-line, and chart, so we—in our travels over the chemical seas—pinned our faith on the immutability of the elements, the atomic weights, and Mendeléef's Periodic Table.

These atomic weights, in which we placed so implicit a trust in those days, were simply figures expressing the relative masses of the unchangeable atoms as estimated by chemical analyses. Thus, if the atom of oxygen weighed 16 units, then the gold atom weighed 197.2, the atom of iron weighed 55.84, and the atom of silver weighed 107.88. We had every reason to rely on the accuracy of these figures, for they had been ascertained by laborious experimental researches made by hundreds of chemists and had been checked and counterchecked again and again by the most refined methods.

It is true that in the historical branch of our subject we had heard the name of Prout—a misguided fellow, we were led to believe—who, in 1816, had put forward the fantastic idea that hydrogen atoms were the basic materials from which the atoms of all the other elements were constructed, just as one can build a shop, a factory, or a villa, by putting the same kind of bricks together according to different plans. A moment's thought was enough to dispose of Prout. In our list of atomic weights, hydrogen's atomic weight was given as 1.008, whilst the atomic weight of lithium was 6.94, which one sees at a glance is not a simple multiple of 1.008. If there were seven hydrogen "bricks" in the lithium atom, then the atomic weight of lithium should be 7.056; if only six "bricks," then the atomic weight of lithium should be 6.048: and neither of those values tallied with the 6.94 which the atomic weight experts told us was the actual value found by experiment. Little did we dream in those days of what was in store; for, to quote the Psalmist: "The stone which the builders refused is become the head stone of the corner," and Prout came, at last, into his heritage after almost exactly a century in the shade.

In 1896 Becquerel discovered radioactivity, which he detected in the case of the element uranium; and, following in his track, Schmidt found that thorium also showed radioactive properties. It is said that a lady once asked Michael Faraday: "But what is the *use* of this latest discovery of yours, Mr. Faraday?" and was met by

Faraday's counter-question: "Madam, what is the use of a new-born baby?" Our knowledge of radioactivity was still under 50 years old in 1945, but even then it had furnished the most destructive weapon within the memory of mankind; and it seems not impossible that its application to industrial purposes may in future make oil and coal seem, by comparison, the mere playthings of a child.

Once the initial discovery had been made, progress was rapid; and by 1903, Rutherford and Soddy had collected ample evidence to support their Disintegration Theory, which accounted for the facts on the simple—but, in those days, staggering—assumption that the atoms of some elements were not stable but were spontaneously breaking down into simpler materials, and that the phenomena of radioactivity were symptoms of this process of decay. In some cases, atoms are surprisingly short-lived. The average life of an atom of the element thoron is only 78 seconds, whilst the average life of a thorium-A atom has been measured and found to be no longer than 0.2 seconds. Thus, by 1903, the old idea of a Universe built up from stable atoms, eternal and immutable, had vanished and the dream of the alchemists seemed to be realised.

At this stage, however, it was only an incomplete fulfilment. The alchemists had aspired to change lead into gold by their own voluntary action, at any moment which suited their convenience; but all that the chemists and physicists of 1903 could do was to watch the progress of a transmutation which they could neither initiate nor control. The breakdown of uranium could not be hastened by making the element red-hot or retarded by plunging it into liquid air; it pursued its course uninfluenced by any alteration in external conditions. Just as a distant observer can watch the progress and measure the speed of an avalanche on an Alpine slope without having the slightest control over its descent, so the scientist could examine the changes in a radioactive material, changes which he was impotent to initiate, to regulate, or to arrest. None the less, the work of Rutherford and Soddy destroyed, once and for all, one of those three fundamental ideas on which the chemistry of my student days had been based. The conception of eternal and immutable atoms had vanished, never to return.

Now take the idea that in a specimen of a pure element, every atom had exactly the same mass as its neighbours. How did this fare in the light of the new knowledge?

It was soon established that when an atom of a radioactive element spontaneously breaks up, it may do so in either of two ways; for the disintegrating atom may either eject an electron or throw out an a-particle.

An electron carries one unit charge of negative electricity; and the mass of an electron was found to be about 1/1,840 of the mass of a hydrogen atom, so that for all practical purposes the addition or subtraction of an electron makes no difference to the weight of an atom, since the change could not be detected by even the most refined chemical measurement. But when an electron is expelled from an atom, the electrical change is easily detectable, since the new material has a unit charge of negative electricity less than the original atom; or, what amounts to the same thing, the material gains one positive charge by the loss of the electron.

Turn now to the a-particle. The investigations of Ramsay and Soddy and the later ones of Rutherford, established its nature beyond doubt. It is an atom of helium, weighing four units, which has lost a pair of electrons and thus has two positive charges.

A rough illustration will make the next stage in the story easy to follow. Most people are familiar with the firework known as a Roman Candle which, when ignited, shoots out some sparks, then a ball of fire, then more sparks, then another fireball, then more sparks, and so on, until the firework's charge has burned itself out and only the paper casing remains. Now suppose we have two different Roman Candles, one weighing 238.2 grammes and containing eight fireballs, each weighing four grammes, whilst the other firework weighs 232.4 grammes and contains six fireballs. For the present purpose we may neglect the weight of the sparks as being too small to come into the problem.*

---

* The sparks represent the ejection of an electron.

Now, obviously, when both fireworks have burned themselves out, the two casings left behind will have different weights. The first one will have lost 32 grammes by expelling its eight fireballs; and the remaining inert casing will weigh (238.2–32) grammes, or 206.2 grammes. The second firework will burn out after losing six fireballs each weighing four grammes, so the weight of the inert casing left behind in this case will be (232.4–24) grammes, or 208.4 grammes.

Bearing this illustration in mind, turn to what happens in the disintegration of the atoms of uranium and of thorium. The uranium atom has an atomic weight of 238.2. It ejects in succession eight a-particles, each weighing four units; so the lead atom which is the final product of the break-down must weigh (238.2–32) units, or 206.2 units. On the other hand, the atomic weight of thorium is 232.4, and the lead atom which is derived from it after the removal of six a-particles must weigh (232.4–24) units, or 208.4 units. But these two forms of lead are found by experiment to be chemically indistinguishable and inseparable. Thus, it is plain that the atoms of the "chemical element" lead may differ from each other by as much as two units in weight; and the assumption that lead atoms are identical with each other in every way vanishes completely from our chemistry.

Soddy tentatively suggested something like this in 1910; and in 1914 he and Hyman produced the first definite experimental proof of the matter. Ceylon thorite is a mineral containing 55 per cent of thorium, 1-2 per cent of uranium, and about 0.4 per cent of lead. This lead is probably the degradation-product of the original thorium and uranium in the mineral, which have been disintegrating through the long geological ages; and since the rate of change of uranium is nearly thrice that of thorium, the lead—if derived from both these elements—should consist of about 10 parts of "thorium-lead" mixed with one part of "uranium lead." An experimental determination of the atomic weight of lead from the Ceylon thorite showed that it was 207.694, whereas the atomic weight of ordinary lead is 207.2. This difference is far greater than any possible experimental error in modern atomic weight determinations. Thus it was definitely established that lead atoms which had precisely the same chemical properties, might differ from each other in weight. Other workers then entered the field, and it was found that, according to the mineral source from which the element was extracted, the atomic weight of lead varied from 206.08 to 207.694; and it became clear that our common lead with an atomic weight of 207.2 was a mere mixture of some of the other varieties, and not a homogeneous material as it had been assumed to be.

The so-called Periodic Law laid down that the physical and chemical properties of the elements are periodic functions of their atomic weights. This simply means that if the elements are arranged in the order of their atomic weights, beginning with the lightest and ending with the heaviest, then similar elements appear at definite intervals in the series, just as octave notes occur at definite intervals on the keyboard of a piano. In 1871, Mendeléef devised a tabular arrangement on this basis; and in his Table every element had a "place"—rather like a square on a chess board or a halma board—and from the "place" in which the element was situated in the Table, it was possible to infer what the properties of that element were.

Naturally, in 1871, certain elements were still undiscovered; and in his Table, Mendeléef left blank "places" to show where these elements should fit in when they were discovered in Nature. Not only so, but from the "places" assigned to these missing elements he foretold with amazing accuracy the chemical and physical properties of these then undiscovered elements. It was one of the boldest prophecies ever risked by anyone in the history of science; but Mendeléef was amply justified by its success.

One of these undiscovered elements was what we now know as germanium. In 1871, Mendeléef predicted that it would be a grey, lustrous metal with a density of about 5.5. Fifteen years later, germanium was actually found in natural sources and it was a grey, lustrous metal with a density of 5.469. Mendeléef's forecast of its atomic weight was almost exact. He put the density of the metal's oxide at about 4.7; actually germanium oxide has a density of 4.703. He even foretold that the tetrachloride of the unknown element would boil at about 90°C. and have a density of about 1.9; germanium's tetrachloride actually boils at 86°C. and has a density of 1.887 at ordinary temperatures. Unfortunately there was a hiatus in the middle of Mendeléef's Table which made it impossible to discover how many elements actually existed in Nature; and the chemist of my younger days was in the position of a sea-captain, groping his way through an intricate archipelago, who picks up his chart only to find that the ship's puppy had got hold of it and eaten a large part out of the middle of the design.

Now, as we have just seen, Soddy had proved that certain materials had identical chemical properties but differed from each other in atomic weight; and a name had to be devised to express this idea. He suggested the word *isotopes* (from the Greek *isos*, equal, and *topes*, a place) to suggest sets of atoms which, though of different weights, are grouped together in a single "place" in the Periodic Table on account of having identical chemical properties. Thus, in the example given above, the atoms of "thorium-lead" weighing 208.4 and those of "uranium-lead" weighing 206.2 would be isotopes of each other.

In 1918 it was my good fortune to recognise a new factor in the atomic weight problem: the fact that two or more atoms may have *identical* atomic weights and yet differ completely from each other in chemical character. Thus, the atoms of the radioactive elements mesothorium-I, mesothorium-II, and radiothorium all have the atomic weight 228, but occupy different "places" in the Periodic Table and have entirely different chemical properties. The reason for this is that mesothorium-I expels an electron in radioactive change and yields mesothorium-II, which immediately picks up an electron from its surroundings, but adds this electron to its system in a place different from that occupied by the expelled electron. Thus, the mesothorium-II atom is different in structure from the parent mesothorium-I atom from which it sprang; but it has precisely the same weight as its parent. I christened these atoms of equal weight and different chemical properties *isobars* (from the Greek *isos*, equal, and *bares*, weight).

Thus, the two types—isotopes and isobars—are in a way the complements of each other. The isotopes are atoms which have *identical* chemical properties but different weights, whilst the isobars are atoms having *different* chemical properties but *identical* weights. After the discoveries of the isotopes and the isobars, it was clear that the old chemical atomic weights were not natural constants but merely the average weights of the atoms in a mixture, and that the atomic weights could be no true guide in the classification of the elements, notwithstanding the success of Mendeléef's predictions.

At this point, the average reader may well say: "This is a very pretty piece of work from the point of view of theory; but what practical importance could it have?" If he will be patient, he will find that the existence of Soddy's isotopes furnished one of the main hindrances to the manufacture of the atomic bomb, whilst my isobars indicated a means of escaping from these drawbacks. The atomic bomb can hardly be brushed aside as of merely theoretical interest.

It will be noticed that the first evidence for the existence of both isotopes and isobars was drawn from the field of the radioactive elements. Would the stable elements also manifest the same peculiarities?

We had not long to wait for an answer. By 1919, Aston had devised and brought into play a wonderfully delicate instrument, the mass-spectrograph. Just as in the ordinary spectroscope, the lines in a spectrum are spread out on a strip, so in the mass-spectrograph the atoms in a mixture were spread out along a strip and from the positions of the marks on the strip it was possible to calculate the mass of the particles which imprinted the marks. Thus, the exact mass of a given atom could be determined; and the accuracy attained was so great that even the weight of the electron, tiny as it is, came within the range.

In an incredibly short time, Aston added more to our exact knowledge of the true masses of atoms than had been achieved by hundreds of workers in three generations. His results showed that, excluding the radio-elements, over 40 of the ordinary elements were mixtures of isotopic atoms, and sometimes very complex mixtures indeed. For example, the familiar metal tin has an "atomic weight" of 118.7; but this is merely the average value of a mixture of 10 isotopes with masses of 112, 114, 115, 116, 117, 118, 119, 120, 122, and 124. As for the isobars, over 30 pairs were detected by Aston among the non-radioactive elements; and there are six cases in which atoms of no less than three elements have the same mass. An instance of this is the triplet argon-potassium-calcium, where the atoms are isobaric and weigh 40 units each, although their chemical properties are completely different in the three elements.

From the point of view of the atomic bomb, the isotopes of hydrogen have special interest. The ordinary hydrogen atom has a mass of 1.008; but there is an isotopic form known which has a mass of 2.0143. This "heavy hydrogen" has been given the special name *deuterium*; and just as ordinary hydrogen combines with oxygen to form water, so deuterium unites with oxygen to yield "heavy water." Ordinary tap-water contains both types of water, but the percentage of heavy water in the mixture is small—about one part in 6,000; and the separation of one variety from the other is a very prolonged and troublesome process. It may be remembered that during the war a special raid was launched on Norway with the object of destroying a German "heavy water" plant which had military importance.

Aston's researches made it necessary to invent a new name which could not be confused with the old "atomic weight"; and he proposed the word "mass-number" to indicate the actual mass of a given atom. When we speak of uranium-235, we mean an atom of uranium which has a mass of 235 units; and silver-107 indicates an atom of silver with a mass of 107 units.

But an even more curious result followed from Aston's investigations. The mass-spectrographic measurements proved, as Aston pointed out, that "the weights of all atoms could be expressed as whole numbers to a high degree of accuracy." Thus, ordinary chlorine, with the "chemical" atomic weight 35.5, actually consists of a mixture of atoms weighing 35 and 37; and the fractional "atomic weight" is merely the mean value of the atoms each of which individually has a mass which can be expressed by a round number. In the background, one sees the shade of Prout.

With the coming of the isotopes and isobars, the old atomic weights ceased to have any physical meaning. But the Periodic Table of Mendeléef still held good as a classification of the elements according to their chemical properties. As the reader will have noticed, that wonderful system had been built up on a wholly erroneous basis—the atomic weights—and yet it had led to triumphs of accurate prediction unsurpassed in the field of science. How can one account for this? With the fall of the atomic weights, the minds of scientists turned to the second factor in the problem of the atom—electricity—in search of something which would account for the accuracy of the Periodic Table.

In 1913, following up hints furnished by the work of Soddy and of Rutherford, van den Broek suggested that the elements should be given "atomic numbers" obtained by counting along the "places" in the Periodic Table, starting with hydrogen. Rutherford had adduced evidence to show that atoms contained a very small "nucleus" in which practically the whole mass of the atom was concentrated; and this nucleus carried a definite positive charge which was neutralised by the charges on a set of electrons which were assumed to circulate round the nucleus like planets round a sun. Van den Broek suggested that his "atomic number" was the number of the neutralising electrons in the atom, or, to put it differently, the number of positive charges on the nucleus. Thus, hydrogen, with one charge on its nucleus and one electron, would have 1 as its atomic number; helium, with two positive charges on its nucleus and two satellite electrons, would be numbered 2, and so on upwards.

Close on the heels of this came the publication of a research which completely altered our outlook on the Periodic Table with its basis in the atomic weights. Seldom has a young scientist—for he was only 26 when he published his discovery—sprung at one leap into the front rank as Moseley did. His single research marked him out as one of the most promising physicists of his time. Promising he remained, for he was fated never to continue his investigations. How little the politicians thought of science in those days may be inferred from the fact that the best use they could find for this brilliant young man was to make him a Second Lieutenant and send him out to Gallipoli, where he was killed in 1915. Fortunately, things are better done nowadays, as the history of Radar shows.

The essence of Moseley's discovery can be seen from a very simple analogy. Imagine a number of bells belonging to a series, though the series is not complete. You are asked to arrange them in order, beginning with those yielding the deeper notes and going on regularly from tenor to treble. You hang up the bells and throw handfuls of gravel at each of them in turn, eliciting the characteristic note of the bell. By measuring the vibration-frequencies of the notes from the bells you would have no difficulty in arranging the bells in proper order; and if any were missing from the series, you could immediately say: "There's a blank in the sequence here, between these two bells."

Now, instead of throwing handfuls of gravel at a set of bells, Moseley bombarded the various elements with streams of high-velocity electrons. In these circumstances, each element emits an X-ray characteristic of the particular element used. By measuring the frequency of that X-ray and comparing it with the frequencies of the X-rays emitted by other elements under bombardment, he was able to arrange the whole series of elements in their natural order; and it was found that the atomic number of the element was the governing factor in the equation which expressed the frequency of the X-ray emitted by that element.

Thus, Moseley's results enable us to make a complete roll-call of the terrestrial elements and to determine that if hydrogen be numbered 1, then uranium is No. 92. Further, the roll-call at once shows not only the existence of any defaulters (hitherto undiscovered elements between hydrogen and uranium) but also their exact position in the series. Thus, we now know how many elements are still to be sought for and what their general character is. The

Periodic Table has in this way been given a logical basis by discarding the atomic weights and replacing them by the atomic numbers, which can be buttressed by the X-ray spectra.

Clearly, if we wish to describe a particular atom definitely, we need to state both its atomic number and its weight. This is done concisely in the following manner: Take, for example, an atom of uranium weighing 235 units and having the atomic number 92. It is completely described by the symbol $_{92}U^{235}$ and if we deal with the mass-factor alone, we can call it an atom of uranium-235.

We have already seen that the pioneers in the field of radioactivity could watch and measure the breakdown of atoms, but could not modify the process in the slightest degree. Ramsay was the first to see further possibilities, and he suggested that the a-particle might be used as an agent to break down stable atoms. The a-particle has a speed of some 10,000 miles per second. Mass for mass, its energy is four million times that of a rifle bullet. Here, then, is a projectile which might be used to break up the stable atoms.

Following this line of thought, Rutherford bombarded some of the lighter atoms with a-particles; and he was able to show that by this method atoms of nitrogen, aluminium, sodium, phosphorus and others could actually be disrupted. More important still, it was found that in these disruptions energy was liberated, and this energy was greater than that of the a-particle which induced the disintegration. In fact, the a-particle was acting like the pull of the trigger by a rifleman which liberates the much greater energy of the cartridge in the breech. Thus, the first step had been taken towards tapping the energy locked up within the structures of the stable atoms.

Strictly speaking, since he was relying upon the a-particle, Rutherford's atomic disintegrations would be described as "controllable" but not "purely artificial"; for the a-particle itself is a product of natural radioactive change. But the next step, taken by Cockroft and Walton in Rutherford's laboratory, resulted in an atomic disruption which was produced entirely by artificial means, without the employment of any natural radioactive substance. By generating between two electrodes a potential difference of some

800,000 volts, they were able to accelerate the speed of ordinary protons—i.e. hydrogen atoms which have lost their electrons and therefore carry one positive charge on each atom—to speeds up to 50,000 miles per second; and when these swiftly-moving particles were allowed to impinge upon lithium, the lithium atoms were disintegrated. Thus, in 1932, the first truly "artificial" break-up of an atom was achieved, since the process is entirely independent of natural radioactive sources.

The machinery of this disintegration of lithium is important from the point of view of the atomic bomb. When the high-velocity proton strikes the lithium atom, it is captured and its mass is added to that of the lithium atom. The resulting agglomeration is unstable and splits up spontaneously into two helium atoms which dart off on different paths. Now the mass of two helium atoms is slightly less than that of a proton plus a lithium atom; so some mass has evidently vanished during the process. What has become of it? According to Einstein, if mass disappears, energy appears in its place; and here, evidently, is the origin of some of the energy which makes the atomic bomb so formidable.*

Cockroft and Walton's results were the first published on the subject of artificial transmutation; but simultaneously in America, Lawrence had been building his cyclotron; and his results, when

---

* The actual figures for the masses involved, as determined by mass-spectra, may be given for the benefit of those interested.

1.008 (Proton) + 7.0161 (Lithium atom) = 8.0241 (Agglomeration)
8.0241 − (2 × 4.0028) (Helium atom) = <u>0.0185</u>.

The underlined figures represent the mass which has vanished and which reappears in the form of energy. According to Einstein, if a gramme of matter is totally annihilated by transformation into energy, the resulting energy will amount to 9×10^{20} ergs. (One foot-pound is equivalent to 1356×10^{4} ergs).

they appeared, revealed that the problem of artificial transmutation had reached the stage of producing quite considerable quantities of material. Lawrence's cyclotron is a machine in which an enormous electro-magnet holds a proton in a plane, whilst acceleration is provided in a most ingenious fashion by means of an ordinary alternating electric current. As a result of the forces applied to it, the proton sweeps round and round in a widening spiral track, making each revolution in the same time as its predecessor. Since the path lengthens at each revolution, the proton has to cover a longer distance each time, and its speed increases until at last it shoots from the discharge tube at enormous velocity and can be hurled at a target composed of the element which it is desired to transmute. The stream of atoms emerging from the discharge tube looks rather like a flame jet. In addition to protons, deuterons† from heavy hydrogen can be used, which are even more shattering since they have double the mass of the proton; and even charged atoms of helium have been employed, which have four times the mass of the proton.

Compared with the cyclotron, the Cockroft and Walton apparatus is like a pop-gun alongside a machine-gun. As early as 1934 Lawrence produced an entirely new radioactive form of sodium by means of his cyclotron. This new "synthetic" form of matter is made up of atoms weighing 24 units, which have an average life of about 15 hours, and break up spontaneously with the formation of an isobar which is chemically identical with magnesium. During their short life they are strongly radioactive, and enough of the material has been prepared to give results equal to those which would be produced by one milligramme of radium. This has already opened up the prospect of "radium treatment" in medicine at far cheaper rates and on a much wider scale.

Thus, at long last the old alchemists' dream of changing one element into another at will has come true. But already science

---

† A deuteron is an atom of deuterium (heavy hydrogen) which has lost an electron and so carries a positive charge of one unit.

has passed that goal in its stride and produced two elements, neptunium and plutonium, which—so far as we know—never existed at all upon earth until they were artificially prepared in our scientific laboratories.

Previous to 1932, all the agents employed in attempts to disintegrate atomic nuclei were atoms carrying positive electric charges. These charges acted as "handles" by means of which the particles could be seized and directed by magnetic and electrical fields; and to this extent they were invaluable. They had, however, one marked drawback. The nuclei of all atoms carry positive charges and the value of the total charge in the nucleus increases from element to element up the series until the nucleus of uranium, for example, bears a concentrated positive charge of no less than 92 units. Now if a proton, carrying one positive charge, approaches the nucleus of a heavy atom, the big positive charge on the heavy atom's nucleus tends to repel it or to divert it from its path; so that instead of hitting the heavy atom's nucleus the proton is pushed aside and passes harmlessly by. This explains why the earlier successes in artificial disintegration were all scored among the lighter elements, such as boron, sodium, aluminium and phosphorus and why, in those days it seemed unlikely that any bombardment by protons, deuterons, or a-particles would succeed even in the case of even such a light element as potassium, the atomic nucleus of which carries a positive charge of 19 units.

It was easy to see that better effects might be obtained by using as a projectile some particle which carried no electrical charge and would not be repelled by the positive charge on the atomic nucleus which was being attacked. But an uncharged particle has no "handle" by which it can be seized, guided, and speeded up by means of electric and magnetic fields, so that road seemed barred.

In 1932, however, came Chadwick's discovery of the neutron, which was the very agent required. Our knowledge of the neutron is still incomplete. It appears to contain a proton and an electron as the hydrogen atom does; but whilst the electron is easily removed from the hydrogen atom, the two charges are so closely associated in the neutron that they are separated only in the violent convulsions of radioactive change. For our present purpose it is sufficient to say that the neutron has unit mass and no detectable electrical charge. It is, therefore, not repelled by the positive charge on the nucleus of any atom which it penetrates and is thus a much more effective atom-smasher than any of its predecessors.

Neutrons cannot, as yet, be obtained directly from the cyclotron or similar instruments, because they lack the electrical charges which serve as "handles" by which they can be gripped. But if the element beryllium is bombarded by high-speed charged particles from the cyclotron, its disintegration yields a supply of neutrons. Once obtained, these neutrons may be used to break up fresh atoms, with the liberation of still more neutrons, and soon. This is what is termed a *chain reaction*.

A simple parallel to the chain reaction will be familiar to anyone who has played dominoes. Set up a series of dominoes on their edges in the order shown in the diagram below:—

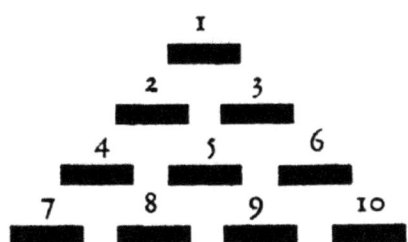

Give No. 1 domino a gentle push with your finger. This represents a neutron entering an atom and upsetting its equilibrium. Domino No. 1 falls forward and upsets dominoes 2 and 3. These, in their turn, fall forward and upset Nos. 4, 5, and 6, and so on. The effect is that the original gentle push has brought down the whole group of dominoes.

Another example of a chain reaction process is the spread of influenza. A child suffering from 'flu goes to school and infects two or three of its schoolmates. Each of these passes germs to several neighbours, and so the disease spreads in ever-widening circles until a serious epidemic is raging in the school.

In theory, the initial problem of the atomic bomb centred in the question: Is it possible to cause "nuclear fission" in a uranium atom in such a way as to start a chain reaction which will produce an ever-increasing supply of neutrons to carry the process forward? If so, then the reaction will go quicker and quicker, involving larger and larger quantities of uranium until explosion ensues. Imagine a neutron let loose in the midst of a mass of impure uranium. Four things may happen: (1) the neutron may traverse the mass and escape out of it without having collided with any uranium atom; (2) the neutron may be captured by an atom of the impurities in the mass; (3) the neutron may strike a uranium atom in such a way that it is captured without producing fission in the uranium nucleus; and (4) the neutron may be captured by a uranium nucleus in such a way as to produce fission in the nucleus, with the liberation of several fresh neutrons. Obviously, the first three processes represent a dead loss of neutrons, whilst the fourth increases the number of available neutrons; and the possibility of the atomic bomb depends on more neutrons being produced in process (4) than are lost in processes (1), (2), and (3).

But, in practice, difficulties arise. Natural uranium is a mixture of three different isotopes, and these are present in very different percentages: Uranium-234 (0.006 per cent), uranium-235 (0.7 per cent), and uranium-238 (99.3 per cent) roughly. Of the three, uranium-235 is the most suitable for the purpose of making the atomic bomb. Both uranium-235 and uranium-238 can capture neutrons, but usually the capture in the case of uranium-238 merely attaches the neutron to the uranium nucleus—producing uranium-239—whereas neutron-capture by uranium-235 produces the nuclear fission required for the liberation of fresh neutrons to carry on the chain reaction. Thus, it was necessary to separate the 0.7 per cent of uranium-235 from the comparatively useless 99.3 per cent of uranium-238. Had this been a chemical problem, no trouble would have arisen; but the two materials are isotopes and inseparable from each other by chemical means. Thus, the properties of isotopes raised an immense difficulty. Instead of an easy chemical separation, physical methods had to be employed, such as a large-scale modification of Aston's mass-spectrograph, gaseous diffusion, magnetic or centrifugal separation, etc., all of which are slow and toilsome.

Even when uranium-235 had been obtained in sufficient quantity, another problem arose. This particular isotope of uranium gives the best results when it is bombarded with slow-moving (so-called thermal) neutrons. Thus, some means had to be devised for putting a brake on the high-velocity neutrons liberated by the fission of uranium-235, so that they might be slowed down sufficiently to attack other uranium-235 nuclei. If the reader will imagine the effect of making a swift-swimming minnow pass from a tank of water to a tank of treacle, he will have a crude idea of the kind of "brake" which is used. This brake is termed a "moderator" and is applied by mixing an impurity with the uranium. Among the most useful moderators are hydrogen and deuterium compounds, beryllium and graphite.

Another factor in the atomic bomb problem was that of the "critical size" of the material employed in the bomb. A very crude illustration will serve here. Suppose that throughout the pulp of an apple, some insect has laid its eggs uniformly. The eggs represent the atoms which are about to break down, liberating neutrons; and the hatching of the eggs may illustrate the actual freeing of the neutrons, which may be pictured as the larvae. Assume that the larvae begin to eat their way through the pulp of the apple. A certain number of them will tunnel out to the rind of the apple and escape; and it is evident that the proportion which can thus escape will depend upon the surface area of the apple. That represents, crudely, the escape of neutrons from the uranium mass. Now if we cut the apple into halves, we expose two fresh surfaces and so increase the area across which the larvae can make their escape; and hence there will be more escapes than before. Clearly, in the case of the atomic bomb, for the sake of keeping the chain-reaction alive, we need to keep the surface area as small as possible in comparison with the number of neutrons within the bomb. Now the quantity of matter in a sphere depends on the cube of the radius, whilst the surface depends on the square of the radius; or, to put it differently, if the bomb be made spherical in shape, the probability of neutron escape depends on the square of the radius of the sphere, whilst the fission effect depends on the cube of the radius. The "critical size" of a spherical bomb is defined as the size for which the production of free neutrons is just balanced by their loss through escape and by non-fission capture within the sphere. If the sphere is smaller than the critical size, then the chain reaction fails, because the production of neutrons by fission does not keep pace with the rate of escape.

The foregoing paragraphs have indicated some—but far from all—of the problems which faced the scientists who devised the uranium atomic bomb. In practice, the greatest difficulty in their path was that arising from the chemical identity of the isotopes uranium-235 and uranium-238;* for owing to this identity, no separation could be achieved by chemical means. Yet compared with the physical methods of separating isotopes, our ordinary chemical methods of isolating one material from another are incomparably easier to employ, when it is possible to use them. Could some other path to the atomic bomb be discovered?

Isobars to the rescue! It was mentioned above that the more plentiful isotope, uranium-238, can capture a neutron without suffering nuclear fission but merely by adding this extra neutron to its structure. This process yields a new, "synthetic" isotope, uranium-239, which is not known in Nature but is a pure laboratory product.†

---

* The nucleus of an atom of uranium-235 contains 92 protons, each carrying one unit positive charge, and 143 neutrons. The nucleus of uranium-238 holds 92 protons and 146 neutrons. In both cases the nucleus is surrounded by 92 electrons spinning in orbits; so the system as a whole is electrically neutral. The atomic number of the element is determined by the number of positive charges on the nucleus, so it is 92 in both of above cases.

† The atom of uranium-239 contains 92 protons and 147 neutrons in its nucleus, plus 92 electrons in orbits round the nucleus. Like the other uranium isotopes, it has the atomic number 92.

Uranium-239 proved to be a fairly short-lived isotope. Its atom spontaneously ejects an electron and changes into an atom of the hitherto unknown element neptunium-239, which in its turn expels an electron and yields an atom of yet another new element, plutonium-239.* And when plutonium was prepared, it was found to be susceptible to nuclear fission just like uranium-235.

What advantages has plutonium over uranium-235 from the point of view of atomic bomb manufacture? In the first place, the raw material is uranium-238, which makes up 99 per cent of ordinary uranium and is, therefore, almost a hundred times as plentiful as uranium-235. Secondly, all the difficulties involved in sifting one isotope from another disappear. Uranium-239, neptunium-239, and plutonium-239 are examples of my isobars and their chemical properties are entirely different; so that from the mixture of them which is produced in the manufacture of plutonium, it is a chemical problem to isolate the pure plutonium—an easier matter than separating two isotopes like uranium-235 and uranium-238, where only physical methods can be used. Thirdly, plutonium had in it the potentiality of a "super-bomb."† But the whole problem of plutonium is at present a matter of official secrecy for security purposes, so it is inadvisable to deal with it further in this place.

Finally, there remains the problem of designing the bomb itself. From what has already been said, the reader will know that so long as the bomb is smaller than the "critical size," it will not explode; but if it is made larger than the critical size, then a chain reaction starts spontaneously and cannot be stopped before detonation occurs. Yet the bomb must be capable of being transported to the target area without risk of premature explosion. How is this to be managed?

The answer suggests itself at once. The bomb can be constructed in two or more parts, each of which is below the critical size; and

---

* See *Atomic Energy*, p. 38. (Published by the U.S.A. Government and reprinted by H.M. Stationery Office, 1945).
† *loc. cit.*

these parts can, at the last moment, be fitted together so that the combination of them exceeds the critical size. For instance, the parts might be made in the form of two equal hemispheres, each below the critical size, and then the two parts might be clapped together, flat to flat, so as to form a sphere larger than the critical size. Each hemisphere by itself would be perfectly safe to handle. But for technical reasons, it is essential that the contact between the parts should be as nearly instantaneous as is practicable; and a better method would be to make the bomb in two parts and then, at the cardinal moment, shoot one part as a projectile into the other portion. But this section of the subject is also secret for security reasons, so it need not be discussed here.

At this stage, it seems interesting to deal with some matters which have accrued from the investigations mentioned in the preceding pages. Space is short and ideas are many; so only a few points can be touched upon here; and the reader must be left, for the rest, to his own speculations.

An entirely fresh chapter has been opened up by the synthesis of the elements neptunium and plutonium. Certainly, by means of Lawrence's cyclotron, new forms of matter had been obtained; but each of them had its place already marked out for it between No. 1 and No. 92 in the Periodic Table, since these new atoms were isotopes of older, well-known types. Radio-sodium was a fresh variety of Element No. 11, but it had all the properties of ordinary sodium with which chemists had been familiar for generations, though it had radioactive properties as well. But the leap beyond uranium to neptunium and plutonium stands in a different category. A method has been discovered of preparing artificially a whole series of more and more complex atoms.

Now, so far as our knowledge goes, the more massive atoms are like castles of cards which tend to collapse spontaneously when their architecture grows over-intricate. Even uranium is rare and if Nature has produced anything more complex, it is only in the minutest quantity. Plutonium may exist in Joachimsthal pitchblende; but it cannot be more plentiful than one part in 10^{14} parts of the ore, or about one ounce in 2,700,000,000 tons. Human ingenuity has far surpassed Nature in this field, for artificial plutonium has been prepared in quantities sufficient for employment in the atomic bomb. Elements Nos. 93 and 94 have been synthesised in the factories, and that opens up a vista along which we can, as yet, only gaze and speculate. Is there any bar to the production of still heavier and more complex atoms? If so, what will be the results when that impassable limit is reached? None can tell at this moment.

Turning to another field, Lawrence's cyclotron has opened up a subject which will take many years to explore. Already it has yielded new artificial radioactive materials, the existence of which suggests to the mind of the chemist wide practical applications. Radiotherapy has added new weapons to its armoury. Another line of work lies among the "tracer elements." Suppose you swallow a dose of sodium phosphate, what happens to it? Even micro-analysis is not delicate enough to follow the phosphorus atoms into all the parts of your organism where they may be playing their parts in the reactions of your bodily machine. But if you mix with your ordinary phosphate, 1 per cent of a phosphate prepared from radioactive phosphorus, then radioactivity measurements, thousands of times more delicate than any chemical determinations, will tell you whither the radio-phosphorus has spread; and since chemically the two varieties are the same, it is clear that for every atom of the radioactive material detected there must be 99 atoms of ordinary phosphorus in the same spot. Thus, as the cyclotron brings within our ken a new set of radioactive forms of ordinary elements, our knowledge of plant and animal metabolism will gain immensely.

When we turn to the many problems raised by the discovery of the atomic bomb itself, the first which faces us is this secrecy question. Is it desirable, or even advisable, that everyone should have presented to him the precise methods of constructing atomic bombs?

Suppose that a score of children have assembled for a nursery tea. One of them has a packet of poison in his pocket. Would any sensible person propose that each of the other babes should immediately be presented with a similar packet of poison so that all may start fair and that if there be a bad boy in the party he should have his chance of poisoning his neighbours? But, say the objectors, within five years all these children are going to be in a position to get hold of a packet of poison in any case, so it makes no difference whether or not you hand it out now. The reply is that a good deal may happen in the course of five years. The children might even (though it is not likely) grow up to the stage of understanding that it is better not to poison other folk.

Those who wish to do so are entitled to draw comfort from the case of poison gas warfare. Hailed as atrocious when first used in 1915, it became a commonplace weapon during the rest of the 1914-18 war, and its employment in 1939-45 was regarded as a practical certainty. Yet it was never utilised on the grand old scale. Possibly history may repeat itself in the matter of the atomic bomb; and its appearance in 1945 may have no sequel in the next war.

It seems doubtful if the use of the atomic bomb would make much difference to military tactics; for once battle is joined, the bomb could hardly strike a foe without injuring a friend. But in the fields of strategy and logistics it would undoubtedly exert a considerable influence. Just as the coming of the machine-gun forced the tactical abandonment of close order fighting, so the atomic bomb would impose the use of a much looser and more complex form of strategy, though the general principles would remain unaltered. What would be the use of gathering together huge masses of men merely to form a convenient target for the atomic bomber? The day of such battles as Verdun, Passchendaele, and the Somme seems to have passed, since the vast gatherings of reserves behind the fighting-line would merely furnish easy prey for the atomic weapon.

As for military objectives, if German ideas prevail, the great cities would be amongst the most obvious, since Hiroshima and Nagasaki have shown how easy it is to reduce the enemy's population by a few strokes. The munition areas are also certain to be devastated if that be possible; and the defence of these will have to be mounted on a scale which will make the London barrage seem trifling by comparison. But a new type of objective would appear with the coming of atomic warfare. Uranium will be of vital importance until it is displaced by some other element, perhaps thorium; and thus, wherever there is a uranium ore deposit a target will be found. Before the last war, no one would have imagined that the Great Bear Lake district in Canada had the slightest military importance; and the Belgian Congo was thought of, if at all, merely as a convenient centre of aerial communications. But if atomic warfare came, places like these and Joachimsthal in Bohemia would rise into vital importance as sources of uranium ore.

Infantry will still remain an essential arm in land-warfare. Conquest is never complete without occupation of the enemy's territory; and in the age of atomic energy, such an occupation will be even more necessary than ever, if merely to ensure that no chance is left to the opponent to manufacture atomic bombs in secrecy and take a careless conqueror at a disadvantage.

No one has yet seen the effects of dropping an atomic bomb in such a way as to penetrate the ground and produce a subterranean explosion. Possibly such a weapon might act as a super "block buster" and shake the foundations of a sizeable city. It might also leave behind it quantities of radioactive matter* which might make the target region dangerous long after the explosion, as H. G. Wells so vividly depicted in his *World Set Free*. If events took this turn, we might see the end of the age of great cities and the reign of Saturn might return, after a period of famine and death.

Soddy† has suggested that the legends of the Garden of Eden, and the Fall enshrine the tradition of an actual far-distant past

---

* "The (U.S.A.) War Department now authorises the further statement that the bomb is detonated in combat at such a height above the ground as to give the maximum blast effect against structures and to disseminate the radio-active products as a cloud. On account of the height of the explosion, practically all the radio-active products are carried upward in the ascending column of hot air and dispersed harmlessly over a wide area." (*Atomic Energy*, p. 127).

† Soddy, *The Interpretation of Radium* (1920), pp. 182-3.

when men attained to the complete control of atomic energy and lived in something akin to the Golden Age, out of which they were precipitated by some unknown catastrophe. I recommend his suggestion to the consideration of those learned divines who meet in conclave from time to time to discuss the atomic bomb from the theological standpoint. It is rather difficult to regard our present-day state as a Golden Age; but a war of atomic energy might well thrust humanity still further down the scale into a condition much more unhappy than the one we live in now.

These are gruesome speculations. Let us turn to a more cheerful picture. Have we the means of controlling the liberation of atomic energy, and, if so, what beneficial results may we reap from this in the future?

As to control, this will demand some method of keeping the chain reaction alive, whilst restraining it within bounds. An ingenious suggestion has been made by Adler and von Halban. In the case of uranium-235, atoms of the element cadmium can capture neutrons produced by the disintegration of uranium-235 atoms; and the rate of capture increases rapidly with the speed of the neutrons which, in turn, varies with the temperature. If some cadmium salt be added to a solution of uranium-235, then the chain reaction will raise the temperature; and this will increase the rate of neutron-capture by the cadmium atoms, so that finally the mixture will reach a constant temperature. If now heat be withdrawn from the system to work machinery, the temperature of the system will fall; the rate of neutron-capture by the cadmium atoms will decrease; the chain reaction will go more rapidly; and again the temperature will rise to the fixed point. Thus, heat can be steadily withdrawn from the system for the production of work by some external machine.

As my colleague, Dr. C. L. Wilson* wrote in 1943, before any news of progress towards the atomic bomb had been made public: "If this almost unbelievably convenient method of control is practicable,

---

* See A. W. Stewart and C. L. Wilson. *Recent Advances in Physical and Inorganic Chemistry* (1944).

then it would seem that the major problem in the production of atomic energy in the future will not be chemical but purely financial, in that it will depend on the economic practicability of obtaining sources of uranium rich in U-135 isotope. Because of the rapidity with which the nuclear field has developed in recent years, there is also another attractive possibility which may be borne in mind. Further investigation in the region of nuclear fission may reveal other, more readily available sources, which will be amenable to similar nuclear chain reactions." If the reader recalls the £500,000,000 spent on what Mr. Churchill called "the greatest scientific gamble in history," and also the discovery and utilisation of plutonium, he will admit that this prophecy was an extremely shrewd one, especially since it was made without any "inside information."

But while we may look forward with some confidence to gaining control of atomic energy, we have in the past a serious warning which we shall be well advised to heed. When the Industrial Revolution came in the nineteenth century, it brought with it the illusion that—since the machine could do the work of a hundred men—working-hours would be shortened and the workman would have far more leisure. In practice, things have not worked out quite in this simple manner; and during the slump which followed the war of 1914-18, we had the fantastic spectacle of part of the population toiling with might and main whilst the remainder of the workers lived miserably on the dole, since there was no employment available to them. If atomic energy is to do much for mankind, this sort of thing will have to be avoided; work will have to be shared out more evenly; and something will have to be done to offer the worker, in his leisure hours, recreations a shade better than the enjoyments at present within his reach, if he should happen to wish for this.

Every new source of power brings in its train an outburst of mechanical ingenuity. During the seventeenth and the first half of the eighteenth century, patents for mechanical inventions were granted at the rate of eight per annum. Then came the steam engine of James Watt in 1769, and it gave stimulus to inventors. From

1763 to 1852, patents for mechanical inventions were taken out at the rate of 250 per annum. In the following two decades the rate rose to 2,000 per annum. By 1877, the annual number had increased to 3,200. Nowadays, with oil and electricity added to steam, the flood of patents has swollen to the extent that in 1941 no less than 11,179 patents issued from the Patent Office. It seems reasonable to assume that the coming of atomic energy will produce an outburst of mechanical ingenuity similar to that which followed the introduction of steam, but on a gigantic scale.

What fresh things are to be expected when atomic energy can be controlled? To begin with a couple of small matters, fresh types of building material may be devised to replace the ferro-concrete of our skyscrapers. Silica occurs in abundance on the earth's surface, and with unlimited heat available, it could be fused into masses of any desired form and size. New types of resistant road-material may also be in store, and we may see Africa and Asia laced with a network of transport arteries.

As for power, it is clear from what has been said above about "critical size," that the likelihood of an atomic engine suitable for transport is small, unless some fresh elements are detected for which the "critical size" is far less than that which suits the cases of uranium and plutonium. Even so, the weight of the shields necessary to protect passengers against the radiations resulting from atomic disintegration would probably make "atomic" vehicles too heavy and clumsy. It seems more probable that atomic energy will be generated at special power-stations from which it can be distributed electrically to places where it is to be utilised.

There is another field in which controlled atomic energy might find an application. It might be employed to remould some of the minor features of our terrestrial globe itself. If it were worth while, no doubt the atomic explosives would make short work of the Kra Isthmus, and cut a sea-level canal which would unite the Indian Ocean to the Gulf of Siam, thus saving the distance through the Malacca Straits. And the project of "flooding the Sahara" (as it was magniloquently described) might easily be achieved. Some 60 years ago, Roudaire proposed to cut a canal from the Gulf of Gabes in Tunisia westward to link up the sea with a succession of depressions in the desert: the Shott-al-Fejej, the Shott Jerid, the Shott Rharsa, and the Shott Melrir, whose northwest end is not far from Biskra. In 1883, the cost of the projected cutting was estimated at £6,000,000. The object was to let the waters of the Mediterranean flow into this depressed area and so produce in North Africa an inland sea with an area of about 3,000 square miles.* The formation of this sheet of water was expected to modify the local climate and make the desert blossom like a rose; but whether it would do so or not is a matter of opinion. Another canal might have more immediate results. With atomic explosives, it might not be a difficult task to drive a sea-level ship-canal from the Bristol Channel into the central plain of England and so bring the teeming industry of Birmingham and its neighbourhood into direct touch with the sea.

Visions on this scale are intoxicating, but one must draw the line somewhere. Even with atomic energy at his disposal, it is unlikely that Jules Verne's President Barbicane could realise his dream of unfreezing the North Pole by altering the terrestrial axis. For one thing, uranium is a rare element, and it is doubtful if there is enough of it in the world to achieve the purpose of the Gun Club.

When radioactivity was a recent discovery, some timorous souls were much disturbed by the fear that an incautious scientist might set up an atomic disintegration which might spread spontaneously from atom to atom until, in Prospero's words, "the cloud-capp'd towers, the gorgeous palaces, the solemn temples, the great globe itself, yea, all which it inherit, shall dissolve." We now know that they may sleep quietly in their beds. As has been shown in the foregoing pages, the interposition of a "moderator" suffices to arrest

---

* That is about fifty per cent. larger than either the Zuyder Zee or the Great Salt Lake in Utah, and thirteen times the size of the Lake of Geneva.

the chain reaction in the atomic bomb; and what are most of the elements composing the terrestrial sphere but a series of moderators? Sleep sound. You are not likely to be wakened by the "Crack of Doom" foreshadowed in Robert Cromie's romance.

IV
HOW TO POISON YOUR INTELLECT

THIS IS NOT AN ESSAY ON PROPAGANDA. Other pens have discussed in detail the advantages and disadvantages of the truth, the half-truth, and the downright lie, considered as means of persuading the simple-minded to believe what they are told in print or by word of mouth; and in the course of two world wars we have had thrust upon our attention enough material of all three kinds to enable us to decide for ourselves which is ultimately the most effective form of argument. The subject to be dealt with in the sequent pages is a much more modest one, to wit, the influence of certain drugs upon the mind of the patient who absorbs them.

Morphia, cocaine, ether, and various other substances have been described as "habit-forming drugs" owing to the fact that indulgence in them produces a craving for further and larger doses, so that finally the victim becomes enslaved to the stimulus and cannot break off the practice. Even nicotine has been classed as a habit-forming drug on the strength of the existence of chain-smokers.

This simplification of the problem, however, leaves out of account the physiological and psychological nature of the partaker, which may have a decisive effect on the matter.* As a pinch of practice is worth a pound of theory, perhaps I may be allowed to quote

---

* This book was read in manuscript form by a medical expert who advanced the criticism that I have not laid sufficient emphasis on "the varying stuff of which intellects are made." "It must not be assumed that because a drug affects

my own experience in this field. When I was a Research Fellow, I used to smoke between forty and fifty cigarettes a day during the University session. This looked as if I had become a slave to nicotine. To satisfy myself that I could break the habit at will, I used to abjure tobacco entirely during vacations until, after a trial of several years, I had convinced myself that I could stop smoking if and when I chose, without feeling the worse for that. Again, in the days before the passing of the Dangerous Drugs Act, when I was in a position of greater freedom and less responsibility than nowadays, I dabbled in experiments on the effects of opium, morphia, cocaine, ether, and other drugs out of curiosity and without acquiring any drug-habits. On the basis of these experiences, it seems obvious that in the formation of a "drug-addict" his personality plays a large part, and that the whole blame cannot be laid on the drug itself.

In considering the psychological effects of these drugs, a difficulty presents itself at the very outset, due in part to the poverty of our descriptive vocabulary and in part to the fugacity of our recollections. It is useless to say, "This flower is yellow if you are speaking to a man who has been blind from birth. The words can convey nothing to him, because he has never experienced the sensation which is implied in them. Further, everyone who has suffered from some peculiarly vivid nightmare knows how fleeting are the details of that dream when he tries to put them into words after his awakening. One would need to be exceptionally gifted in order to convey the atmosphere of terror and horror which bathes one during the original vision. Probably the same is true in the case of drug-generated phantasies and illusions, and this may account for the difficulty encountered when an attempt is made to determine their precise details by questioning the person who has

> (cont.) one person in a certain manner it will have a similar effect on another person. The varying effects of alcohol, the most widely employed drug in the Western Hemisphere, are proof to the contrary." I shall be glad if the reader will bear this in mind.

experienced them. In practice, it is rare that he can do more than indicate broad outlines.

Among the material poisons of the intellect, opium has gained the widest publicity. This advertisement is due in part to the widespread use made of it in the East, to the Opium War against China, and to the literary fame which it attained through the works of Baudelaire, De Quincey, Coleridge, and others. As most people know, it can be enjoyed by smoking or eating the dried juice of a particular kind of poppy; or by drinking its tincture, laudanum; or by swallowing or injecting hypodermically the principal alkaloid in opium, morphine; or else by snuffing up heroin, a diacetyl-morphine prepared in the laboratory from morphine itself. Of these methods, opium smoking is probably the least harmful, since the weight of alkaloids conveyed in the smoke and absorbed from it by the body must be very minute as compared with what can be injected with a hypodermic syringe or swallowed.

According to Lewin*, the opium addict passes through certain stages in his career. In the first stage, he suffers from delusions which magnify to himself the value of his intellect, his work, and his agreeable sensations. He lives in a world of his own, cushioned off from the minor frictions which beset him in ordinary life. Then, as the dose is increased during months, a second stage ensues, during which the addict is filled with a feeling of complete contentment, well-being, freedom from desire, and mental calm. The field of interest is contracted till only the addict's own ego seems of importance, and a moral deterioration sets in as a result. In this phase, an addict may ignore social responsibilities and his craving for the drug grows so acute that he will steal, if that be necessary, in order to procure the next dose. Then another stage is entered, in which willpower sinks and expires. It is impossible to face even the most every-day situations without the reinforcement supplied by ever-increasing doses of the drug; and yet every dose now causes degeneration in the physical machine. Nutrition is disorganised,

---

* Lewin, *Drugs* (1938), pp. 58-60.

emaciation sets in, headaches grow chronic, and all care for personal hygiene is lost. In the latest stage, the morphinist is hardly more than skin, bone and rasped nerves.

A couple of instances of the lengths to which addicts will go in pursuit of morphine have fallen under my own observation and may be given here. One of these unfortunate creatures found himself unable to procure a supply of his drug and consequently, no doubt, suffered considerably from this forced abstention. But his ingenuity triumphed temporarily over the difficulty. He dragged himself through the streets and when he saw the brass plate of a physician, he fell upon the doorstep, writhing and crying aloud in simulated agony. After attracting attention, he explained that he suffered from angina pectoris, or some other excruciating ailment, and begged for a hypodermic injection of morphia to soothe his pangs. By such tactics he sometimes succeeded in persuading the doctor to "give him a jag," whereupon he betook himself a little further, until another brass plate set him to his acting again in the hope of a fresh dose. This procedure could hardly be prolonged indefinitely. Word went round the medical fraternity and his little play ceased to be effective.

The other case affected me more directly. A morphomaniac managed to find his way into my private office one day whilst I was lecturing; and by rifling my desk, he succeeded in stealing some official printed order forms. On these, he forged orders for a supply of his drug; and by taking them to some dealer he hoped that these faked documents would get him what he wanted. Unfortunately for himself, he was unaware that the forms which he stole were only for intra-University use, and were invalid if sent direct to suppliers; and this led to his capture and consignment to some curative institution which, I hope, weaned him from his infirmity. So much for the moral collapse which awaits a weakling who is foolish enough to fall under the influence of morphia.

De Quincey, vague as his descriptions are, has given us the best accounts of the visions generated by opium. In Part III of *The Confessions of an English Opium Eater*, he mentions a phenomenon of childhood which I remember experiencing myself, and which is no doubt common enough. "I know not whether my reader is aware that many children have a power of painting, as it were, upon the darkness all sorts of phantoms: in some that power is simply a mechanic affection-of the eye; others have a voluntary or semi-voluntary power to dismiss or summon such phantoms." For the sake of clarity, I think I should add that—in my own case, during childhood, at any rate—one has the actual impression of "seeing with the eye" in such cases, and not merely of conjuring up mental phantasmagoria in one's brain. In De Quincey, addiction to opium re-awakened in adult life this faculty which most of us lose with the passing of our childhood, but coupled it with something else. "At night, when I lay awake in bed, vast processions moved along continually in mournful pomp; friezes of never-ending stories. . . . And concurrently with this, a corresponding change took place in my dreams; a theatre seemed suddenly opened and lighted up within my brain, which presented spectacles of more than earthly splendour."

There is nothing uncommon in the co-existence of a "seen" world and a "dream" cosmorama. Any moderately good sailor can recall lying in his bunk during rough weather and consciously watching the swaying of a curtain or the swinging of a door-hook whilst synchronously the mind ran in reverie and explored Thule; or how the clink of a tumbler in its rack, or the noise of footsteps in the companion-way accompanied, but never mingled with, a chase after the unicorn through moonlit forests, or some vast conflagration in China.

In *Their Lawful Occasions* Kipling depicts the simultaneous impressions of two alien worlds which came to him as he lay in his bunk in Torpedo-boat No. 267. "Anon I caught the tramp of armies afoot, the hum of crowded cities awaiting the event, the single sob of a woman, and dry roaring of wild beasts. . . . Through all these phenomena and more . . . my shut eyes saw the lamp swinging in its gimbals, the irregular gliding patch of light on the steel ladder, and every elastic shadow in the corners of the frail angle-irons; while my body strove to accommodate itself to the infernal vibration of the machine." Here it is noteworthy that Kipling writes of his "shut eyes," indicating that his visual impressions were not those of normal sight but were, like De Quincey's, a phantasm painted upon darkness.

The difference between the "normal" and the "opiumised" state appears when we find that with De Quincey, "whatsoever I happened to call up and trace by a voluntary act upon the darkness was very apt to transfer itself to my dreams," whereas in the normal experience the impressions from the real world will no more mix with the dream than oil will coalesce with water. The two exist side by side at the same moment, independent and unblending.

Another characteristic of opium's effect noted by De Quincey is the alteration of the sense of space and, at a later stage, of the sense of time. Everything appeared on a titanic scale. "Space swelled, and was amplified to an extent of unutterable and self-repeating infinity. . . . Buildings, landscapes, etc., were exhibited in proportions so vast as the bodily eye is not fitted to receive. . . . I beheld such pomp of cities and palaces as never yet was beheld by the waking eye, unless in the clouds . . . Cities and temples beyond the art of Phidias and Praxiteles, beyond the splendours of Babylon and Hekatompylos. . . . The mountains were raised to more than Alpine height, and there was interspace far larger between them of savannahs and forest lawns; the hedges were rich with white roses; and no living creature was to be seen, except that in the green churchyard there were cattle tranquilly reposing. . . ."

In this last sentence appears another tendency of the opium dream. Normally, the dreamer is an actor, and often the principal actor, in a drama which unrolls about him during his sleep. Rarely indeed does he play the part of a disembodied spectator, watching from without a piece in which he has no part. I must have had thousands of dreams myself, but among them I can recall only a single example in which, as a viewless ghost, I watched the crumbling and collapse of some unheard-of civilisation. But under the influence of opium, the reverse is the case, and it is the "personal" dream which is exceptional, whilst the "spectatorial" one is common. If humanity finds its way into these visions, it is almost invariably generalised and blended into the anatomy of some titanic picture.

Read that masterpiece of De Quincey's rhetoric, the description of the dream from which he awoke in struggles and cried aloud, "I will sleep no more!" What did he experience? "The feeling of a multitudinous movement, of infinite cavalcades filing off, and the tread of innumerable armies. . . . Sudden alarms; hurryings to and fro; trepidation of innumerable fugitives . . . and, at last, with the sense that all was lost, female forms, and the features that were worth all the world to me; and but a moment allowed—and clasped hands, with heart-breaking partings, and then—everlasting farewells!" Those familiar with the original passage will concede that in it De Quincey far surpassed almost all his competitors in transmuting the essence of a dream into cold print. Yet there is not a single individual figure in his picture.

The same peculiarity appears in other visions. "To my architecture succeeded dreams of lakes and silvery expanses of water. . . . The waters gradually changed their character—from translucent lakes, shining like mirrors, they became seas and oceans. . . . Now it was that upon the rocking waters of the ocean the human face began to reveal itself; the sea appeared paved with innumerable faces, upturned to the heavens; faces, imploring, wrathful, despairing; faces that surged upwards by thousands, by myriads, by generations." Here, also, the vision is of masses, not of individual faces.

Even when a recognisable human figure does appear, it seems inconsistent with the opium-engendered atmosphere. No one can doubt De Quincey's affection for Ann, whom he lost in the mazes of London; for it is the most poignant feeling expressed in the whole of his *Confessions*. "To this hour I have never heard a syllable about her. This, amongst such troubles as most men meet with in this life, has been my heaviest affliction." Yet when he meets her again in his opium-dream this is what happens: "Not a bow-shot from me, upon a stone, shaded by Judean palms, there sat a woman; and I looked, and it was—Ann! She fixed her eyes upon me earnestly; and I said to her at length, 'So, then, I have found you at last.' I waited; but she answered me not a word." And straightway the whole pageantry of the opium background fades outright, "and in the twinkling of an eye I was far away from mountains, and by lamplight in London, walking again with Ann—just as we had walked, when both children, eighteen years before, along the endless terraces of Oxford Street." Plainly, then, the intrusion of a powerful human emotion into the dream sufficed to shatter the whole fabric, to dissolve those "domes and cupolas of a mighty city" and to send Thomas back to pace the terraces of Oxford Street by night and tread again the pavements of that "stony-hearted stepmother" from which he had fled so many years before.

It might be suggested that this "ab-human" attitude of De Quincey in the majority of his dreams was a mere personal proclivity, and that the effects of opium upon other personalities might be entirely different. But this is not so. In Baudelaire, the same "ab-human" tendency is present; but in him it attains an extreme in which humanity, in any shape or form, is wholly banished from the vision. Read his *Rêve Parisien* in which the opium influence is more marked than in any other of his poems. It depicts him wandering amid a labyrinth of colossal architecture akin to that described by De Quincey. Here are two of the stanzas:—

> "Babel d'escaliers et d'arcades,
> Cétait un palais infini,
> Plein de bassins et de cascades
> Tombant dans l'or mat ou bruni;
> Et des cataractes pesantes,
> Comme des rideaux de crystal,
> Se suspendaient, éblouissantes,
> A des murailles de métal."

No life of any sort, not even a flower, in this world of metal, marble, and water, where gigantic naiads mirror themselves in sleeping pools girded by colonnades, and channels of blue water run between quays of rose and green, away to the very limits of the universe. Obviously one might seek for humanity through all the squares and arcades of this titanic "folly," but seek in vain.

As a third piece of evidence, perhaps I may be permitted to mention the effects of my own restricted experience of opium smoking. I found myself wandering through titanic halls, with massive galleries running tier above tier to vertiginous heights, whilst higher still hung the traceries of the vaulted roofs, far-distant yet visibly gigantic beyond anything compassable by human endeavour. Imagine a gnat crawling on the pavement of St. Paul's and able to appreciate the scale of the building. Tiny figures moved hither and thither, lost in that immensity of stone; but I met none of them at hand. The only emotion which I recall was a faint disquiet, an uneasy feeling that this builded desolation belonged to Something with which I had no kinship, and which might resent my trespass. But even that impression was faint and formless.

In dealing with the effects of opium, it is impossible to omit Coleridge from consideration. Even the most casual reader of the three great poems can perceive that they fall into two classes: *The Ancient Mariner* and *Christabel* belong to one group whilst *Kubla Khan* stands apart from them. The two former poems have definite plots and depict personalities; *Kubla Khan*, so far as it was completed, is almost purely descriptive, a "spectatorial" fragment like the dream-pictures of De Quincey.

Now it is fortunate that we have the key to this difference furnished by Coleridge himself in his foreword to *Kubla Khan*. He tells us that "in consequence of a slight indisposition, an anodyne had been prescribed, from the effect of which he fell asleep in his chair at the moment that he was reading the following sentence, or words of the same substance, in "Purchas's Pilgrimage": "Here the Khan Kubla commanded a palace to be built, and a stately garden thereunto; and thus ten miles of fertile ground were inclosed with a wall." The author continued for about three hours in a profound sleep, at least of the external senses, during which time he had the most vivid confidence that he could not have composed less than two to three hundred lines; if that indeed can be called composition in which all the images rose up before him as things, with a parallel production of the correspondent expressions, without any sensation or consciousness of effort. On awaking, he appeared to himself to have a distinct recollection of the whole, and taking his pen, ink, and paper, instantly and eagerly wrote down the lines that are here preserved."

Clearly, if for the euphemism "an anodyne," we read "laudanum," we have an account of an opium-dream growing out of the quotation from Purchas which was in Coleridge's last waking thoughts. Then appeared that regrettable "person on business from Porlock"; and by the time his affair had been settled, the remainder of *Kubla Khan* had vanished from the memory of Coleridge, never to return. But in the fragment which Coleridge completed, we have what is probably the most accurate reproduction extant of an opium vision in detail.

Cocaine may be dismissed more briefly. Originally, the coca tree was found only in South America, but now it has been introduced into Java, Ceylon, and India. The alkaloid, cocaine, is obtained from coca tree leaves. It is valuable as a local anæsthetic by its action on the terminations of the sensory nerves, though it is often replaced by novocaine (procaine) and other synthetic substitutes; but from our present point of view it is important owing to its action on the central nervous system.

Medical authorities are unanimous in condemning it as a "habit-forming" drug; but there is distinct disagreement among them over any possible advantages it may have. It seems beyond dispute that for centuries the natives of Peru and Bolivia have used it to increase their resistance to physical fatigue and to hunger; but probably this last effect is due simply to its anæsthetic action on the nerves of the stomach.

With regard to its mental effects, it seems to produce excitement, restlessness, and garrulity at first, but on this follows a phase of calm and languor. It banishes depression and fatigue. Some people ascribe to it the quality of enabling partakers to carry out long spells of intense mental activity without much rest; and other even more surprising properties have been described. From my own experiments with it, I found that it produced a sense of wellbeing, a certain relaxing of mental strain, and a tendency to sit quietly in an arm-chair without bothering much about anything. As a mental stimulant it proved valueless in my case. In fact, it made me kin to the old countryman who described his activities: "Sometimes I sits and thinks, other times I just sits." And I am inclined to believe Cushny's verdict*: "Some travellers in South America relate marvellous tales of its producing feelings of the highest bliss and power, but these have not been confirmed by experience of cocaine in less romantic regions of the globe."

In addicts it eventually gives rise to sleeplessness, digestive disturbances, hallucinations, convulsions, delirium, and insanity—a price which seems rather high to pay for any pleasant sensations one may derive from the drug. One of the hallucinations is characteristic of cocaine: a feeling as if there were grains of sand (or worms and insects crawling about) under the skin.

Another drug which attacks the central nervous system is found in hemp plants (*Cannabis indica* or *Cannabis sativa*) from which are obtained the preparations known as bhang, hashish, charas, ganga, and kif. It is used in various ways: by smoking it either alone or mixed with tobacco; by making it into sweetmeats; in the form of an intoxicating drink; or by infusing it with coffee. It must be one of the most widely-employed among the "habit-forming drugs," for it has been calculated that over a hundred million people make use of it.

Though it came into Western medicine little over a century ago, it has been used in Asia and Africa so long that the origin of the practice is lost. Herodotus (IV: 73, 75) notes how the Scythian crept into tents and threw hempseeds upon red-hot stones. "Immediately it smokes, and gives out such a vapour as no Grecian vapour-bath can exceed; the Scyths, delighted, shout for joy, and this vapour serves them instead of a water-bath; for they never by any chance wash their bodies with water."

The effects of hemp seem to vary considerably from individual to individual and from race to race, so that only a general account of its action can be given here. Initially, sensations of restlessness and anxiety may appear, but they soon pass into a feeling of happiness. Judgement is upset and imagination is untrammeled by reason. In the Eastern races, the dreams of the cannabist have an

---

* Cushny, *Pharmacology and Therapeutics* (1936). p. 443.

erotic tinge; the European may laugh, sing, or make ridiculous gestures. Sometimes dual consciousness appears; and usually the ideas of time and space become confused. Ordinary sounds seem magnified into thunder; rockets and stars may be seen; and the well-known "flying dream" is experienced.

Alternating with the pleasure, however, are other illusions of an afflicting character. Mortal fear attacks the cannabist and he shivers in terror of imaginary dangers.

In the long run, the addict seems to pay dearly for his pleasures. Sense-delusions become terrible; persecution-mania may appear. In one very numerous group, every excess of hashish leads to maniacal excitement and aggressive violence of a dangerous kind. In not a few cases, the addict sinks into permanent imbecility.

A picture of the delusions and the childish indecency produced by bhang is to be found in the *Arabian Nights*.* A poor fisherman, when his day's work is done, eats his daily dose of bhang and spends the evening talking to himself. Late on one fine night, he decides to take a walk down to the river; but when he gets to his doorstep the bhang deludes him into mistaking the moonlit square for his fishing ground, and the shadow of the walls for its bank. He gets his tackle, baits his hook, and casts it out into the supposed river, whereupon it is seized by a passing big dog. Bemused by his drug, the Bhang-eater mistakes the cur for a huge fish; and when, in its struggles, it drags him to the edge of the shadow which he takes for the river-bank and threatens to pull him over into what he imagines to be the river, he cries for help. Neighbours come at his call and try to persuade him that he is deluded; but he refuses to listen to them, so they laugh at him and go away, whilst he continues to cry for aid. Later, the Kazi rides that way, accompanied by his

---

* See, in Sir R. F. Burton's translation of the *Book of a Thousand Nights and a Night*, "The tale of the Kazi and the Bhang-Eater," "The History of the Bhang-Eater and his Wife," "The Tale of the Hashish-Eater," and Burton's footnotes on bhang and hashish.

guard, and the Bhang-eater appeals to him lest the monster fish should drag him into the river and drown him. The Kazi's attendants unhook the dog and the Kazi, recognising the man as a bhang-eater, and being himself an addict, carries him off to his own dwelling, where, next night, the two sit down to enjoy a debauch on the drug. Presently they become intoxicated: the Bhang-eater dreams himself to be the Sultan, whilst the Kazi promotes himself to be Governor; and they discuss the affairs of the city.

Meanwhile, the Sultan and his Wazir in disguise come to the door and, hearing all this tumult, enter the house, whereupon they are drawn into the discussion and finally the Bhang-eater "Sultan" orders his "Governor" to strike off the real Sultan's head. The Sultan is furious, but the Wazir explains the state of affairs. "These twain are eaters of Hashish, which drug when swallowed by man garreth him prattle of whatso he pleaseth and chooseth, making him now a Sultan then a Wazir and then a merchant, the while it seemeth to him that the world is in the hollow of his hand."

The Sultan is curious, and next night, after the Kazi and the Bhang-eater have taken their normal dose, the Sultan and the Wazir return and offer them more bhang, which they eat. "Now when they had taken an overdose, they got into a hurly-burly of words and fell to saying things which can neither be intended nor indited, and amongst these they exclaimed: 'By Allah, the Sultan is deposed and we will rule in his stead.'" Whereupon the Kazi and the Bhang-eater attempt to treat the real Sultan and his Wazir as Gulliver treated the Queen's palace.

On the following night, the Bhang-eater and the Kazi sit down once more to their drug. Presently their heads whirl round, when their drunkenness says to them, "Up with you and dance." When the Sultan and the Wazir appear, the pair keep on dancing until they are tired and are compelled to sit down and take their rest.

As can be seen, this tale contains all the main features of bhang's effects: the delusions about the physical world, the violent itch for talking, the delirium of grandeur, the Rabelaisian indecency, and the uncontrollable inclination towards violent exercise.

Much milder in its effects than hemp is the drug known as kat or gat. It comes from the bush *Catha edulis* (*Celastrus edulis*), which is cultivated in cool valleys at an altitude of three or four thousand feet in south-west Arabia and north-east Africa. Usually it is enjoyed by chewing the young buds and fresh leaves of the plant. Aden has a special kat market; and as the plant will not grow in low lying country, bundles of it are brought down by express messengers during the night.

In some ways, kat resembles the coca plant, since it revives energy, keeps down the pangs of hunger, and makes the eater wakeful. Like the popular cocktail, it is a social drug which promotes friendly intercourse. It produces a pleasant excitement and gaiety; and at "kat-parties" the kat-eater rejoices when he hears other people talking, and he himself tries to contribute to the general entertainment. It is not clear, however, whether it improves the conversation or whether, like the cocktail, it merely stimulates a dull dog into volubility without making him any the less a dull dog.

Like all the other drugs which attack the central nervous system, kat has its defects when used in excess. The kat-eater loses his sleep and grows restless. Chronic loss of appetite and heart-trouble are apt to supervene. It also appears to diminish inclination for the other sex; and inveterate kat-eaters seem to remain bachelors.

One of the most interesting of all these poisons of the intellect is derived from *Anhalonium lewinii*.* The upper part of this plant consists of grey-brown lumps, roughly circular in form, about 1½ inches in diameter, crowned with a dirty-white woolly cushion. These are called "mescal buttons"; but the plant has many names among the tribes who use it. It is found only in restricted areas on the dry plateaux of northern Mexico, remote and difficult of access. It came to the notice of Europeans after the conquest of Mexico by Cortez, but it had been used by the Mexicans for centuries before that, under the name *peyotl*.

---

* For the facts given here I am indebted to Lewin, *Drugs* (1938).

Modern knowledge of this cactus dates from 1886, when some specimens came into the hands of Lewin during his travels in America. Examination showed that the plant contains several allied alkaloids of which the chief one, mescaline, has a structure of the same type as the well-known substance adrenaline. Mescaline appears to be the root-cause of the hallucinations produced by "mescal buttons"; but the other alkaloids seem to have a modifying influence upon the illusions.

According to Lewin—and his verdict is borne out by much evidence—"this anhalonium towers above the rest of the plants on account of the special character of its effects on man. No other plant brings about such marvellous functional modifications of the brain." Indeed, in early times; the Mexicans venerated it as a god, a vegetable incarnation of some divinity. It was considered by them as sacred, whereas the Roman Catholic missionaries regarded it as a work of the Devil, and tried to stamp out its use. It was employed by the natives in religious ceremonies, the cactus being either chewed and eaten or else stewed to yield a hallucinating liquor.

Perhaps the best way of conveying the impressions produced by mescal is to quote portions of a very complete description given by Dr. Beringer* of the hallucinations experienced by him when under the influence of mescaline.

"My ideas of space were very unusual. I could see myself from head to foot, as well as the sofa on which I was lying. All else was nothing, absolutely empty space. I was on a solitary island, floating in the ether. No part of my body was subject to the laws of gravitation. On the other side of the vacuum—the room seemed to be unlimited in space—extremely fantastic figures appeared. . . . I saw endless passages with beautiful pointed arches, delightfully coloured arabesques, grotesque decorations. . . . These visions changed in waves and billows, were

---

* Beringer, *Experimentelle Psychosen durch Mescaline.* Vortrag auf der südwestdeutschen Psychiater-Versammlung in Erlangen, 1922.

built, destroyed, and appeared again in endless variations, first in one plane and then in three dimensions, at last disappearing in infinity. The sofa-island disappeared . . . I was seized with passionate curiosity, great things were about to be unveiled before me. I would perceive the essence of all things, the problems of creation would be unravelled. . . . Then the dark room once more. The visions of fantastic architecture again took hold of me . . . Crystals appeared again and again, changing in form and colour. . . . Then the pictures grew more steady, and slowly two immense cosmic systems were created. . . . Shining with their own light, they appeared in unlimited space . . . The systems approached each other and were attracted and repelled. . . . I saw two cosmic systems . . . in perpetual combat. . . At the beginning they moved at a giddy speed, which gradually changed to a quiet rhythm . . . The moment drew near when both polar systems would be able to oscillate together, when their nuclei would combine in a tremendous construction. Then everything would become visible to my eyes. I would experience everything, understand all. . . . A disagreeable trismus tore me away in this moment from the supreme tension. . . . I felt great unhappiness and profound discontent."

Again and again, just as he felt on the point of gaining full comprehension, there came the attack of lock-jaw, the gritting of his teeth rousing him at the crucial moment. "I was not to penetrate the mystery. I was standing in the midst of the evolution of the universe. I experienced cosmic life just before its solution. The impossibility of understanding the end, this refusal of knowledge, was exasperating. I was tired and felt bodily suffering."

The effects of mescaline generally last for about six hours. The visual hallucinations occur in the dark or when the eyes are closed, but they are noted also in other circumstances. Things assume such brilliant colours that by comparison the real world seems merely drab; and the colours themselves are of a delicacy and variety such as are never seen in normal life. Rather rarer are aural illusions of wonderful music far away. Sometimes perfumes are perceived. Depersonalisation may occur or a dual personality may arise. Yet through all this the subject preserves his normal concentration of thought, and may be quite conscious of the fact that he drugged himself.

The peculiar interest of mescaline lies in the fact that it forms a bridge between two well-defined types of drug. On the one hand lie those substances, like opium, hashish, etc., which yield visions that pass, pageant-like, through the mind without exciting the reason to inquire into them. On the other side are chemicals such as ether and chloroform, which give the impression of keying the intelligence to abnormal pitch without any concurrent pictorial fantasies. As is plain from the foregoing account, mescal has a foot in both camps: it conjures up opium-visions of marvelous architecture, but it also sets the mind to work upon intricate problems suggested by the imagery which it creates. A point of special interest is this peculiar feeling of standing on the brink of the secret of the Universe, which Dr. Beringer describes, for it reappears even more markedly in the case of ether and chloroform.

Many people have had experience of ether and chloroform when undergoing surgical operations; and possibly some of these may find it difficult to recognise the phenomena which are described in the following paragraphs. "We never felt anything like that," they may object. Probably this is so; but the explanation is not far to seek. In anesthesia with ether there are at least two stages, a borderland state and a condition of complete insensibility in which thought ceases; and it is in the borderland phase that these peculiar phenomena occur. The object of the professional anesthetist is to reduce his patient to complete insensibility as quickly as possible, consistent with safety; and therefore, in his technique, the borderland is crossed so quickly that the patient is hardly aware that he is traversing it. The investigator of these peculiar sensations, on the contrary, seeks to prolong the borderland phase and avoid the transition into total insensibility.

A clear description of the effects of ether is given by de Maupassant. When suffering from an attack of migraine, he inhaled ether to dull the pain; and he has recorded his experiences.*

---

* See de Maupassant, *Sur l'Eau*, pp. 98-101 (Edition Conard, 1921); and compare *Rêves* in *Oeuvres Posthumes*, Vol. I, pp. 64-65 (Edition Conard, 1929).

"After a few minutes, I heard a vague murmur which soon passed into a sort of humming, and I felt the whole interior of my body growing light, light as air, vaporising. . . .

"Soon, the strange and fascinating sensation of emptiness which I felt in my chest spread further, reached my limbs, which in their turn grew light, light as though the flesh and bones had dissolved away and left only the skin. . . . And I heard voices, four voices, two dialogues, without understanding the words. At times there were only indistinct noises, anon a word reached me. But I knew that this was merely the accentuated humming in my ears. I was not sleeping, I was awake, I understood, I was quite conscious, I reasoned with amazing clearness, profundity, intensity, and with an intellectual joy, a curious intoxication arising from this ten-fold increase in my mental powers.

"It was not a dream such as hashish yields, nor was it like the rather sickly visions of opium; it was a prodigious acuity of the reasoning power, a new way of seeing, judging, appreciating things and life, with the certitude, the absolute knowledge, that this new way was the true one.

"And the old phrase of the Bible flashed into my thoughts. It seemed to me that I had tasted of the tree of knowledge; that all mysteries were being unveiled, so much was I under the sway of a new, strange, and irrefutable logic. And arguments, ratiocinations, proofs came in throngs, only to be overturned immediately by some stronger proof, objection, or argument. My brain became an arena in which ideas battled with each other. I was a superior being, armed with an invincible intellect, and I tasted a prodigious enjoyment in proving my powers . . ."

An even more vivid impression of the effect of ether is due to Sir William Ramsay. With his usual ingenuity in surmounting practical difficulties, he broke away from all the errors which unavoidably creep in through the inaccuracy of our after-recollections in such cases, by employing a stenographer to take down his descriptions of his sensations while he was actually in the "borderland" condition. Anyone familiar with the phenomena of ether narcosis at this stage will recognise at once how accurately Ramsay reproduced the mental peculiarities of that state, though perhaps to the non-initiate it may not be so clear.

Perhaps, from my own experience, I may be able to clarify some of de Maupassant's description.

So far as I am concerned, ether produces no flood of hallucinations such as come from opium. The only one I can recall—and it occurred but once—is a feeling that I was passing through a maze of tattered, grey, filmy, cobweb-like stuff, rather disagreeable to see and touch, which lined the labyrinth in which I was wandering. But it hardly interested me, so intent was I on the thoughts which occupied my attention.

The mental effects were peculiar. As a disembodied intelligence, I seemed to have hit upon an entirely fresh mode of reasoning, which is extremely difficult to describe or even indicate. If the succession of our thoughts during ordinary life be represented by an arithmetical progression, 1, 2, 3, 4, 5, etc. in which the reasoning proceeds deliberately and securely from stage to stage, the ether-governed intellect thinks in a sort of geometrical progression, 1, 2, 4, 8, 16, 32 . . . which is just as logical, but poles asunder from normal speculation. In other words, during everyday life our cogitation moves step by step, pedestrian-wise; whilst under the influence of ether the mind seems to spring from island to island across streams of thought. Yet despite this advance by leaps and bounds, the logic appears at the moment to be far more convincing, and the inferences attained seem more irrefutable than anything experienced during our normal reflections.

Nor is this all. These giant strides from one thought to another lend to the operation a semblance of swiftness which makes the normal process of cerebration appear to go at a mere snail's pace by comparison. This, in all probability, contributes to the feeling of augmented intellectual power which de Maupassant describes.

In that "ether-borderland," humanity and human affairs vanish completely from the field, leaving only purely intellectual interests in play. I became akin to some super-mathematician following a train of thought towards the solution of a problem which had no touch with gross matter in any shape or form. I was drawn onward, forging link after massive link in a logical chain, but ceasing to care for the links already joined. They had served their purpose, and the new ones seemed so much more interesting and important. Always I felt that just ahead lay a complete and satisfactory solution, and that when it was attained, the whole meaning of the Universe would become lucid as crystal. Finally, I reached the very brink. One step more, just one, and everything would be clear. And at that point—invariably—I came out of the narcosis with a feeling of defeat and acute disappointment. If only one had been allowed that final step! But one never is.

Ether-drinking was a popular vice among the peasantry of the North of Ireland in the latter half of the nineteenth century. Two things about this have puzzled me. In the first place, how did they manage to swallow the stuff? Even though they used water as a "chaser," the effect on the mucous membrane cannot have been pleasant. Secondly, what did they experience in their potations? I can hardly believe that the yokels of the North of Ireland were deeply interested in the secret of the Universe. One is driven to the conclusion that they sought something else, something suggested by the dictum: "Sure, ayther's grand stuff! You can get dhrunk on it thrice in twenty minutes!" But times have changed. Ether is fallen, and in its stead there reigned for a time that unhallowed concoction "Red Biddy."

What has been said above with reference to ether holds good for chloroform also. The illusions produced by the two drugs are similar; but chloroform is practically immiscible with water, whilst ether is soluble to some extent, and this probably has some slight influence upon the actions of the two substances when in contact with the blood.

Another hallucinatory material, nitrous oxide, used to be commonly employed in dental practice, but its use seems to be on the decline. One of its early investigators, Sir Humphrey Davy, inhaled fifteen quarts of the gas, after which: "I danced about the laboratory as a madman." Another inquirer said that after breathing the gas: "I burst into a violent fit of laughter and capered about the room without having the power of restraining myself." From effects of this kind, the gas was nicknamed "Laughing Gas." My only personal experience of its effects has been at the hands of dentists, and possibly the environment was not conducive to laughter. Certainly I felt no inclination towards that. But the true explanation probably is that, as in the case of ether, hallucination occurs only in the "borderland" stage of anesthesia, and in dental technique this phase is traversed so quickly that the patient hardly recognises it before he passes into complete insensibility.

In conclusion, a word of caution seems called for. If anyone feels an inclination to experiment with habit-forming drugs, he should recall *Punch's* advice to those about to marry: "Don't!" If you feel a craving to poison your intellect, then adopt the easiest and handiest method. Seize on the first passing specimen of one-sided propaganda and swallow it whole. You will always, if you are that sort of person, be able to find the antidote by absorbing an equivalent of propaganda on the opposite side. You will be a fool for your pains, no doubt, but not such a fool as you would be if you tampered with habit-forming drugs.

V
MONEY FOR NOTHING

IN "A TALE OF A TUB," Swift defines happiness as "A Perpetual Possession of being Well Deceived," and later on he writes of "the Serene Peaceful State of being a Fool among Knaves." If the Dean speaks truth, then most of us should enjoy a considerable share of felicity whenever we venture off our well-known beaten tracks. No one is easier to swindle than the smart business man when he wanders into unfamiliar fields; few people are more readily cheated than the scientist who pits himself against the fraudulent medium on that medium's own territory. In most of such cases, the thing is almost a "certainty" for the trickster. The successful *chevalier d'industrie* is no fool; he spends his time and concentrates his whole attention upon details of the trick which he proposes to employ. Against this, what chance has a person whose interests lie in an entirely different field and who is prepared to devote only a little of his leisure to the problem of detecting the swindle? In many cases he is probably unaware that a trap is being laid for him, and he walks into it precisely as the deviser intended him to do.

Some people, I notice—for they are generally voluble on the subject—object to betting on the ground that, as a result of a wager, money passes from one person to another without any "work done" for the benefit of society as a whole. I am surprised to find, however, that these people are ready to accept a legacy or to draw a war loan dividend, in return for which they have contributed no work whatever. If I have long ceased to "put a bit" on a horse, it is due not to the arguments of these zealots but to the realisation that (in the words of some unknown sage) "betting is a mug's game."

Even if one ceases to bet, however, one may still take an interest in the theory of the business. For example, what is a fair bet? From the point of view of sportsmanship, certain things must be barred; and among them is special knowledge which makes the bet a certainty. For example, if by accident or malice aforethought I overlook my opponent's hand at poker and note that he has a couple of pairs whilst I hold a straight flush, I should hardly be justified as a sportsman in taking advantage of that knowledge to "skin" my adversary. Still less desirable would be the use of reflectors, hold-outs, marked cards, and sleight-of-hand in order to jog Fortune's elbow at a critical moment and persuade her to favour me.

But things are not always so clean-cut as this. Suppose that a punter has—or imagines he has—some "inside knowledge" which inclines him to favour one particular horse in a race: is he entitled, as a clean sportsman, to put his money on that horse? I must leave this to the judgement of the reader. My own view is that no one is entitled to bet on a certainty, but I never knew a Derby (or any other decently-conducted race) which could be regarded as a certainty for one particular horse beforehand. "Horses are not machines," as someone remarked; and even the best horse may be off-colour on the day of the race.

As the late R. A. Proctor* pointed out, even the most honest sportsman may, through ignorance, offer wholly unfair odds. For instance, Lord Yarborough used to offer 1,000 to 1 against the chance that the hand to be dealt would be a "Yarborough"—i.e., a whist hand containing no card higher than a nine. The odds look generous enough. Stake £1, and if the next hand is a Yarborough you pocket £1,000. Actually, though quite unwittingly, Lord Yarborough was putting himself on velvet; for if the odds be calculated mathematically from the theory of probabilities, he should have offered £1,828 to £1 in order to be strictly fair.

---

* R. A. Proctor, *Leisure Readings*, p. 275 (1896).

There is one point to be noted, however. The mathematical calculation is based on the assumption that a very large number of trials is made, in which case a Yarborough hand should turn up once, and once only, in every 1,828 deals. But, obviously, that Yarborough might appear at the first deal and be followed by 1,827 non-Yarboroughs.

One sometimes sees curious things at the bridge table. (I speak of the hands, not of the players or their methods.) As an example, I may mention an experience of my own which I have never seen duplicated. It happened in the days before Auction came in, and at that time the dealer declared trumps after examining his cards. Having dealt, I looked at my hand, found only black cards in it, and went spades. The first card was played, whereupon my partner put down a dummy without a single red card in it. In other words, my partner and I held all the clubs and spades, whilst our opponents had nothing but hearts and diamonds. Naturally, I put down my cards and claimed a grand slam, since I could take the first trick with a trump and after that our opponents could neither follow suit nor trump. Perhaps I may add that this occurred in the middle of an evening so that the cards had been well shuffled before the deal, and my modest sleight-of-hand is quite incapable of arranging such a result, even if I wished to do so.

But let us return to the subject of "money for nothing."

Proctor* tells the tale of a bookmaker who, by some means or other, had learned that if there are ten horses in a race, they might come in (apart from dead-heats) in no fewer than 3,628,800 different ways.† Armed with this "certainty," the bookie decided to turn it to account by inviting his friends to guess the correct figure

---

* R. A. Proctor, *op. cit.* p. 276.

† It requires no abstruse knowledge to check this. Any one of the ten horses may come in first. Suppose that one of them does so. Then for the second arrival we have nine horses to choose from, any one of which may be selected. Thus there are ninety possible ways in which the first two

and to back their guess by a bet. Judging from the results, one Simple Simon apparently jeered at large figures. "In how many ways can ten horses be placed?" quoth he. "Under a thousand, I bet." Confronted with the figure 3,628,800, he scoffed and refused to pay. The question was referred to Proctor for an expert decision; and much to the surprise and vexation of the bookie, he supported Simple Simon, who had been cute enough to re-word the original proposition before making his bet. Ten horses can "come in in 3,628,800 different ways; but only the first three horses are "placed" in technical language. Now out of ten horses we can select 120 different sets of three; and each set can be placed in six different ways:—

1st Place	A	A	B	B	C	C
2nd Place	B	C	A	C	A	B
3rd Place	C	B	C	A	B	A

so there are only 120 x 6 ways in which horses can be "placed" out of a field of ten, or 720 possibilities in all. Simple Simon had not been so simple as he seemed. The would-be sharper had been sharped himself; but he was quick to see a silver lining. "Ah, well," he remarked philosophically, "I shall win more out of this little trick—now I see through it—than I lose this time."

While on the subject of bookmakers, let us examine the risk they run in connection with starting price betting. By arrangement, a bookmaker will receive bets at the starting price by post, or by telegraph, or by telephone, provided that the message has been despatched by the sender before the name of the winner has been

(cont.) positions might be filled. The third horse must be chosen from the eight remaining after the first and second have been selected; so that for the first three horses there are 10 x 9 x 8 possible arrangements. Thus all the possible ways in which ten horses can come home are covered by the figures 10 x 9 x 8 x 7 x 6 x 5 x 4 x 3 x 2 x 1 = 3,628,800.

announced on the course. Therefore if a crook wishes to swindle a bookmaker, he must manage to despatch his message *after* the winner's name has been given out, and yet do so in such a way that the message itself bears plain evidence that it was sent off *before* the race. To most people, the hour in the postmark of a letter or the hour written on a telegram form would seem good enough proof in a case of this kind.

The earliest method employed by crooks is as follows. The swindler puts a blank sheet of paper into a stamped envelope and addresses the envelope *in pencil* to himself, either at his own abode or at an accommodation address. He posts this production so that it will be delivered shortly before the race is run; and the postmark can be quoted as evidence of this time of delivery. As soon as the envelope comes back to him, the crook steams it open, extracts the blank sheet of paper and replaces it by a letter making a bet on the winner of the race, news of which has just come over the wires. He then rubs out the penciled address on the envelope and writes instead in ink the address of the bookmaker whom he proposes to victimise. By these means, the crook now has ready a letter which has apparently passed through the post before the race was run. All he has to do is to stroll to the bookmaker's office and drop his epistle into the letter-box, whence it will be extracted, later, amongst the rest of the mail

This system was too good to last. Bookmakers realised what was happening; and their counter to the method was to discard letter-boxes at their offices and insist on the postman bringing the letters in personally Consequently, the crook found himself faced with the problem of getting his fraudulent letter into the hands of the postman and having it delivered along with the rest. How is this to be achieved?

What happens is this. The crook watches for the postman, walks up behind him and drops the faked letter on the pavement. He then walks on ahead of the postman to avoid any suspicion. Meanwhile a confederate on the other side of the street calls "Hi!" to the postman and points out the letter. "You've dropped that one." The postman innocently picks it up, adds it to his bundle, and carries out the delivery, so that the fraudulent bet comes in with the rest of the mail.*

Turn now to roulette. It is many years since last I played, but I confess to a kindly feeling for this particular form of gambling. As practised at Monte Carlo, it is fair, it is simple, and (as they advertise in hairdressers' shops where the staff exceeds the demand) there is no waiting. One knows one's fate almost as soon as one has put one's money down. And perhaps my kindliness is further influenced by the fact that I have generally been fortunate with any small bets that I have made in this field.

The late Professor Karl Pearson,† after conducting an elaborate statistical examination of the results of over 32,000 spins of the wheels at Monte Carlo, reached the sage conclusion that Monegasque roulette is not a game of pure chance. When I read his book, I was not much impressed by the conclusion since I had already reached it for myself by a less laborious reasoning process. Only a perfect machine could yield results strictly according with the laws of probability, and no perfect machine has ever yet been produced, nor are we likely to see one in the future. Even the finest chronometer deviates to a greater or less extent from accuracy; and the wheels at Monte Carlo are obviously not likely to beat chronometers in precision. Professor Pearson's discovery left me cold, and added nothing to my information.

In the days when gold was one of the currency metals, an ingenious gentleman devised a method of making "easy money" at the tables. At the very moment when the croupier announced the result of a spin, this genius put a five-franc piece on the winning number. Naturally he was requested to remove it, as he had been too late in staking. He took away his silver coin, but left behind on

---

* If the reader is interested in other methods of the same kind, an account of them is to be found in G-*Men of the G.P.O.*, by Mr. F. C. Cartwright, late Chief Investigator of the Investigation Branch of the General Post Office.

† Karl Pearson, *The Chances of Death*, Vol. I, p. 42 (1897).

the table a gold twenty-franc piece which he had concealed under the larger silver coin. A confederate claimed the gold piece as his stake, and the bank paid him—at least until the authorities grew suspicious.

Everybody knows that, owing to the presence of zero on the wheel, the bank has an advantage over the player. Not so many people, however, seem to notice that it is a double advantage. There are thirty-seven numbers in all, including zero. Suppose you put £1 on a number. You have one chance of winning against thirty-six chances of losing; therefore, if the odds were absolutely fair, you should receive thirty-six times your stake if your number turns up. Actually, however, the bank pays you only thirty-five times your stake. That is the first advantage. The second advantage lies, of course, in the fact that if zero turns up, you lose your stake. Most people are content with seeing the obvious, and they conclude that the bank makes its profits on the zero and remain oblivious of the fact that, in addition, the bank is offering them rather short odds.

If you wish to study advantages to the banker in their finest bloom, I recommend an examination of the gambling wheels employed at country fairs. Here, usually, the squeezing of the odds is so remarkable that the bank needs no zero to help it. Suppose that one-half of the disc were painted white and the other half black. If you stake on white you would obviously have an even chance of winning. If there are three equal sectors on the wheel, the odds against you would be two to one; if there were four sectors, the fair odds would be three to one, and so on. In the case where there are nine equal sectors on the wheel, the odds are eight to one against you. The proprietor must live (though, personally, *je n'en vois pas la nécessité*); so he gives the odds an affectionate squeeze and offers, at most, seven to one for that particular chance, or possibly even five to one. Now if all the sectors were of equal size, the greediness of this would be perceptible at a glance; so in practice the sectors are irregular in dimensions, though the principle remains the same. It would need a very good eye to estimate the proper, fair odds on a particular sector on one of these wheels. One may be fairly sure, however, that the punter is not the person who gets the advantage. The same applies to the spinning disc at which the competitor throws a dart.

So much on the question of odds, assuming that the wheel spins uncontrolled. But this is not always the case. Mechanism has been invented which enables the croupier, after a look at the stakes, to enlarge or contract the openings into which the ball must make its way as it settles down, thus favouring one set of numbers over the rest. The contrivance is too intricate for description here, but readers may consult the account given by J. N. Maskelyne.*

In the Black Museum at New Scotland Yard, I was once shown a contrivance which for ingenuity, simplicity, and audacity could hardly be surpassed as a means of cheating. The base of it was an ordinary small roulette wheel with the usual 37 compartments painted red and black and numbered. Now this machine was meant for use at country fairs, where it was placed on a table in the open air; and under such conditions, since the ball was a small one, a gust of wind might have blown it clean off the wheel and so spoiled that turn. To guard against such an accident, a removable tin shield had been provided, in shape rather like a very large inverted egg-cup or a diabolo, with one cone only half the size of the other, both cones being hollow and connected by a small hole at their junction. This "extinguisher" fitted on to the wheel, covering the numbered compartments and leaving a small outer rim exposed. During the spin, wheel and cover rotated together. The mouth of the larger cone was fitted with a metal diaphragm, soldered in position and pierced, close to the periphery, with 37 holes just large enough to allow the ball to slip through. When the cone was in position on the wheel, each of these holes lay directly over one of the numbered compartments Both the "extinguisher" and the visible margin of the wheel were painted with a crude fantastic pattern, such as one used to see on tin tea-caddies or the old tea-chests.

"This is how it works," explained the detective who was acting as cicerone.

* J. N. Maskelyne, *Sharps and Flats*, pp. 269 ff. (1907).

He placed the "extinguisher" carelessly on the wheel and gave the latter a spin; next he threw the ball into the upper cone, where it swirled round for a moment, then vanished through the hole into the larger cone, where it would land on the metal diaphragm and eventually find its way through one of the orifices into a numbered compartment.

"It will be No. 17, this time," said the detective.

He stopped the wheel, lifted the "extinguisher," and, sure enough, the ball had landed in No. 17 compartment. The detective recovered it, replaced the "extinguisher," started the wheel spinning, and handed me the ball.

"You can throw it in yourself, if you like."

I did so, whereupon my guide said:

"This time it will be in No. 10."

And, sure enough, it was there when the "extinguisher" was removed. The detective replaced the "extinguisher," handed me the ball.

"Try again. It will be No. 33 this time."

Various further trials left me still bemused. The same number never turned up twice, but at each test my guide could predict the result beforehand.

"Can you make 21 turn up?" I asked, giving a random figure.

"Oh, yes."

And No. 21 duly appeared.

Anyone can appreciate the advantage which control like this gives to the operator. He need never pay on a number which has been heavily staked; and he can always arrange that a confederate appears to win heavily, thus encouraging dupes.

The explanation? It is so simple that one feels vexation at not having detected it immediately. The ball, after disappearing through the hole between the two cones, does not fall on the diaphragm. Instead, it runs down a pipe to one particular orifice in the diaphragm, and so into the compartment which lies below that hole. All the remaining thirty-six holes in the diaphragm are "duds," so far as the ball is concerned. The painted pattern on the "extinguisher" indicates to the initiate the position of the tube; and all that is necessary is to place the "extinguisher" on the wheel so that the tube comes over the number which is required to turn up.

It is not often that the *Oxford English Dictionary* errs, but it did so when it defined "Martingale," as applied to gambling, in the words: "Gambling system of doubling stakes in hope of eventual turn of luck." The ancient method of "double or quits" is far from being the only martingale extant. For the sake of simplicity, let us confine attention to gambling upon even chances, such as the toss of a coin or the appearance of red or black in dealing from a pack of cards.

The "double or quits" system has the very modest aim of ensuring that the first winning coup after a series of losses will cancel the total loss and make, in addition, a profit equal to the original stake. Thus, suppose that one loses the following stakes in three successive coups: 1, 2, 4, the total loss incurred is 7 units. The next stake is 8, and if one wins on that coup, one secures not only the seven units lost, but the gain of the original stake of one unit. This holds good, of course, no matter how many losses are made before a win occurs.

A more ambitious martingale is that in which each successive stake is double the previous one, plus one unit: 1, 3, 7, 15. . . . Here an eventual win not only recoups all previous losses but ensures a gain of one unit for each coup which has been played. Thus assume that the stakes 1, 3, and 7 have been lost, and the stake of 15 wins; then eleven points have been lost and fifteen have been won, making a profit of four points on the four coups played, or one point per coup.

In the Paroli martingale, the gambler plays for two successive wins, leaving his original stake plus his winnings, if any, on the table for the ensuing turn.

The D'Alembert martingale, sometimes called the *montant-et-descendant* system, consists in adding one unit to the stake after a loss and deducting one unit from it after a win. Thus if a player starts with a stake of five units and wins, his next stake will be four units. If he loses this, the following stake will be five units; and if he again loses, he must increase his next stake to six units, and so on.

The Labouchère martingale is best understood by using a concrete example. Suppose the player has successively lost the stakes, 1, 2, 3, 4, 5, 6, 7, which he has written down on paper as shown. To find the amount of his next stake, he adds together the two figures at the extreme ends of the line (1 + 7) and puts eight units on the table. If he loses, he writes 8 at the end of his line of figures. If, on the other hand, he happens to win, he scores out the two extreme figures, thus: ~~1~~, 2, 3, 4, 5, 6, ~~7~~, and makes his next stake equal to the sum of the two extreme uncanceled figures (2 + 6) = 8. If he wins again, he erases the 2 and the 6, thus: ~~1~~, ~~2~~, 3, 4, 5, ~~6~~, ~~7~~, and makes his next stake the sum of the two extreme uncanceled figures (3 + 5) = 8, as before.

At first sight, martingales seem to ensure that the punter is bound to win if he goes on playing long enough; but in practice he is limited by the extent of his available capital and by the bank fixing a maximum stake. Suppose, for example, that a man has £5 to risk in tossing "heads or tails," that he starts with an initial stake of one shilling and that he doubles his stake each time he loses. If he loses six times in succession, his loss will then amount to 63s. His next stake should be 64s., but only 37s. of his original £5 remains to him, so he will not be able to table the necessary stake and will end up with a loss of three guineas. If he were betting on the "even chances" at Monte Carlo, a similar result would be produced by the fixing of the maximum stake which he is not allowed to exceed.

In any gambling transaction, the philosopher will say to himself: "I propose to lose a certain sum of money without injuring myself seriously. Which martingale should I use in order to prolong the pleasure of gambling and let me play the greatest possible number of coups before this capital disappears? "Only the fool will imagine that any gambling system" provides a sure road to fortune. He is in the same boat as the old faro-player whose dying words of advice to his friends were: "Always copper the Queen on the last turn."

Enough has been said now about betting; let us turn to a fresh field. In his essay on *Diddling Considered as One of the Exact*

Sciences, Edgar Allan Poe distinguishes between the "diddler" and the "financier." "Your diddler is minute. His operations are on a small scale. . . . Should he be tempted into magnificent speculation, he then, at once, loses his distinctive features and becomes what we term 'financier.' This latter word conveys the diddling idea in every respect except that of magnitude." Since Poe did not disdain to describe the diddler, neither need we; but a single example will suffice.

A woman goes into a fairly busy shop, makes a small purchase, tenders a ten-shilling note in payment, takes her change, and leaves the premises. Very soon, however, she comes back. "I find you've given me too little change. I paid you a £1 note and you seem to have mistaken it for a 10s. one. I didn't notice at the time that the change was 10s. short." Possibly the shopkeeper pays, thinking he has made an error. If he refuses to pay, the dame produces a letter. "I got this to-day from my brother. You see from it that he sent me a £1 note for a birthday present. The note was pinned to the letter—there's the mark of the pin. He mentions the number of the note: Z 10 D 249458. That was the note I paid over the counter a few minutes ago. Would you mind looking in your cash register, please? You'll find I'm right." The shopkeeper does so; and, sure enough, he finds that particular note amongst his other takings; so he naturally assumes that he must have made a mistake and he pays over the extra 10s.

How is it done? Very simply, as in all other diddles. A confederate of the woman called at the shop earlier in the day, made a trifling purchase, and paid for it with the £1 note numbered Z 10 D 249458, which thus found its way into the till, ready to be claimed by the woman as hers.

In earlier days, I used to receive pathetic letters asking me to advance money for the purpose of bribing gaolers to facilitate the escape of an innocent man immured in a Spanish prison. Sometimes, to heighten the colour, the prisoner had a beauteous daughter in distress. Virtue was not to be its own reward in this case. The Spanish prisoner, alone of all men, knew the location of a sizeable treasure; and if I would finance his escape, he would recover the treasure and share it with me. Having always steeled my heart against the woes of the Spanish prisoner and his lovely daughter, I cannot speak from experience about the next stage in the affair; but I suspect that it would take the form of a demand for more money to cover unexpected expenses during the escape of the poor prisoner. In recent years I have not come across these frantic appeals from Spain. Possibly the Spanish prisoner has died in captivity or, again, the swindle no longer catches the gulls. But it certainly had a long run, and must have netted a fair income to its promoters while it lasted.

The field of chemistry furnishes a happy hunting-ground for the swindler. Few business men know even the pot-hooks of the subject, and, unless they take expert advice, they fall easy victims to the plausible shark who has got up the elements of a chemical vocabulary.

Once upon a time a wonderfully-efficient and amazingly quick sugar-refining process was hawked about the City. It was not a mere affair on paper; for there was an actual factory working for the benefit of possible investors. It was a most imposing, two storeyed erection, with the refining machinery carefully boarded up, because it was necessary to keep the secret of the process, pending a patent. It was really necessary to keep the secret, as it was such a simple one. The refining was done on the top flat and the sugar came down through a closed chute to the ground floor, where it was put up into hogsheads. Crude sugar was poured into the machinery above, and pure refined stuff came out of the end of the pipe in a very short time. It would have been amazing if it had been a continuous pipe; but it was not. The crude sugar went off by one channel; refined sugar came in by another; and the discontinuity was concealed by the floor through which the piping passed. It looked quite continuous enough for selling purposes, however, and that was sufficient from the vendor's point of view.

A number of years ago, I was consulted by a business man, by no means a fool, in fact considerably above the average in brains and certainly successful in his own line of work. It appeared that he had been approached by an inventor who had discovered a process for synthesising platinum by an economical method. My business acquaintance was quite satisfied about the genuineness of the claim and was prepared to put money into the affair on the strength of his own judgement. At the last moment, however, he decided to have his opinion strengthened by an expert confirmation, and he brought the matter to me.

The man in the street is not familiar with platinum in bulk, but this is a common sight to the chemist, who is forced to use platinum dishes and crucibles for certain operations. The most cursory inspection of the specimen of "synthetic platinum" so proudly exhibited to me by my acquaintance was enough to show that, whatever it was, it was not platinum. One had merely to bend it between one's fingers to know that.

"Leave this stuff with me for examination. I'll let you have my opinion to-morrow."

When an electric spark is passed between two electrodes of a metal, a spectrum is emitted which can be recorded by means of a spectrograph. Each metal has its own characteristic series of lines which appear in the photograph. If the specimen under examination contains two metals, the lines characteristic of each metal appear on the plate. Thus it is a simple, though rather laborious, task to determine what a metallic specimen contains. I examined the spectrum of this "synthetic platinum." It contained not a single line characteristic of platinum; but the photograph showed all the lines yielded by some of the commoner metals. The "synthetic platinum" was, in fact, merely one of various non-rustable alloys of which, even in those days, numerous kinds were on the market, and which the reader probably knows through his table-knives or the heads of his golfing irons, mashies, and niblicks.

I passed this information to my acquaintance. Possibly it dissuaded him from investing capital in the "synthesis" of platinum. Perhaps it failed to convince him, however; in which case the inventor must have obtained money for nothing, or at least in return for a combination of audacity and plausibility. It was no affair of mine, and I heard nothing further on the subject.

Another inventor asserted that he had achieved the following feat. He began by electrolysing a pound of water, thus obtaining a mixture of the gases oxygen and hydrogen. Some of this mixture he tapped off and stored. The remainder he used to drive an explosion motor which turned a dynamo generating electricity. This electricity he used to electrolyse a second pound of water. Thus each run yielded a surplus of the oxy-hydrogen mixture, which could be used in a second explosion motor to yield power. And so on.

When this scheme was submitted to me, I felt strongly inclined to laugh, but one must never laugh in business affairs. So to clarify the matter, I inquired: "I suppose you see that your inventor is claiming to have produced perpetual motion? In fact, he seems to have gone one better than that, for not only does he keep his perpetual motion machine running, but he gets energy out of it as well."

"But it might be possible," my client contended. "Why not?"

"It's your money, not mine," I admitted; and I explained, in the simplest possible language available to me, just "why not."

"Of course, if you think that your inventive friend has made a discovery which completely upsets the work of Faraday, Joule, and Kelvin, then that's your affair. They're generally supposed to have been geniuses in their line."

But it is hard to uproot a fixed idea from the mind of anyone, and the inventor had evidently made a profound impression on his client.

I never heard the upshot of this affair, either.

Another common bait thrown out by the flat-catcher is "a secret process." This sounds mysterious and saves a sharper from entering into any troublesome explanations. I was once consulted by a friend about some wonderful process which had been offered to him, by which a valuable material could be synthesised. Naturally, I asked to see a specimen. This was in the days when casein products and formaldehyde plastics were in their infancy.

"If it's all right," my friend assured me, "you can come in on the ground floor when the company's formed."

I have but little enthusiasm for such ground floors, unless I know something about the cellarage; so I examined the specimen. It was a sample of a type of synthetic material then coming into the market.

"I'd like to see the patent specification covering this stuff," I said. "There *is* a patent, of course?"

"Oh, no! It's made by a secret process!"

I fear I was unmoved by this assurance.

"Do you know how it's made?"

"No! *It's a secret process* I tell you!"

"H'm! But suppose that, after you start manufacturing, some other company comes down on you for infringing its patents, what are you going to do about that?"

Apparently this aspect of the matter had not struck him. A secret process was a secret process, the ultimate word in mystery. No money of mine went into the projected company, and I have never seen anything to make me regret my abstention. How the matter ended, I do not know. That is what happens so often in these affairs. It is like reading an instalment of a magazine serial story and never hearing the beginning or the end of the yarn. One is left, faintly amused, to speculate on how it began and how it all ended.

If anyone desires to set up as a casuist and seeks material on which to exercise his talent, I recommend some examples from the field of finance; for, undoubtedly, it is sometimes difficult to estimate the precise degree of moral responsibility borne by a principal. Further, although the details of fraudulent schemes are often extremely complicated, the schemes themselves are generally based upon some very simple general idea.

Take, for instance, the following. On January 1, shares in—let us say—Consolidated Cocoanuts are standing at £3. A philanthropist begins to advertise broadcast that by June 1 the price of these shares will reach £5. Some small investors, attracted by the confident tone of the advertisement, begin to buy Consolidated Cocoanuts; and, as a result of their demands, the price of the shares begins to rise. By the end of the month it has reached, say, £3 10s.

Further advertisements follow, pointing to this rise as a proof that the original statement was correct. These advertisements attract more small investors; the demand for Consolidated Cocoanuts increases; their price rises, and so on. Eventually the prophecy is fulfilled and the shares touch £5. Up to that point, a case might be made out for the philanthropist. He has spent his own money lavishly on advertisements for the good of the small investor, who now holds shares quoted at £5 which he bought at a lower price, say £4. It is true that these shares are not worth more than £3; but we may leave this point to our casuist for examination.

So much for what appears on the surface. But suppose that our philanthropist, before issuing his advertisements, has quietly bought large numbers of Consolidated Cocoanut shares at £3. He, as well as the small investor, now owns shares which are quoted at a price much higher than that which he paid for them. Now suppose that, as soon as the price of the shares reaches £5 this philanthropic gentleman throws on the market all his holding in Consolidated Cocoanuts.

It is for the casuist to say whether this departs from honesty. Surely a man is entitled to sell his own property for the current market price. And, after all, the labourer is worthy of his hire. He has spent his money on advertisements. Why should he, as a consequence, be debarred from profiting from the rise in price, like any other speculator?

Of course, as a result of his flooding the market with his shares, the price will be forced down; but there is a good margin between £5 and £3; and probably our philanthropist will come out with a considerable profit, even if the shares eventually fall to the original price. The small investor is the person who pays the piper in a ramp of this kind, since he soon finds Consolidated Cocoanuts quoted at a price lower than the one he paid when he bought them. But is the small investor a subject for pity? After all, he was simply trying to get money for nothing, and if it ends in his getting nothing for his money he is hardly in a position to blame anyone but himself.

A more difficult case for the casuist was that of Alfred William Carpenter and the notorious Charing Cross Bank, which collapsed in 1910. Carpenter was, from my experience of him, a silly and fussy old man with no great intelligence and a whole-hearted confidence in his own acumen which was quite unjustified. These characteristics made him an easy prey for anyone seeking a financial backing for enterprises which might have been turned down at the first glance by a more sceptical mind. One example from my personal knowledge will suffice to portray him. I was called in as a consultant to examine a process in which Carpenter had complete faith. This process had been patented in two different countries, so, naturally, I procured the two patent specifications and compared them. In one specification it was laid down that, during the run of the process, the temperature must be kept below 60 deg. C., whereas the other specification insisted that the temperature must be kept above 80 deg. C. Any schoolboy could see at once that one or other (or both) must be incorrect. Thus either Carpenter had proposed to invest a large sum—I heard £50,000 mentioned at one time—without even looking at the patents; or else he was too stupid to read the specifications with any intelligence. None the less, it was obvious to anyone who met him that he regarded himself as a financial genius with an infallible flair for a good investment.

Carpenter began life as a clerk in an accountant's office; but after three years of this, in 1870 he joined hands with a fellow-clerk and set up as an accountant and partnership agent. Six months later the concern was converted into a moneylending business under the name of The Charing Cross Advance and Deposit Bank; and in 1878 Carpenter, having parted from his associate, continued single-handed. In 1886, the title of the business was altered to The Charing Cross Bank, *tout court*. In this avatar the moneylending side of the concern was continued, but a fresh line of activity was begun in addition.

Seductively-worded advertisements appeared in many newspapers, offering fantastic rates of interest upon deposits in The Charing Cross Bank: 6 per cent. if the money was left for six months, 7 per cent. if it was deposited for a year and so on, up to 10 per cent. if the deposit was a five-year one. Naturally, no one accustomed to dealing with money would put his spare cash into such a concern. It was from the uneducated poor that Carpenter got the bulk of his clients. Dazzled by the high rates of interest offered, too inexperienced to understand the risks, they thronged with their little savings to the doors of this "Bank" which promised them so much. Who would be content with the mere 2½ per cent. offered by the Post Office Savings Bank when, by going round the corner into Bedford Street one could get 6 per cent. on one's money? To them, no doubt, it was "money for nothing," and they threw into Carpenter's coffers not mere thousands, but hundreds of thousands. When the crash came, it involved these trustful creatures in ruin.

Paradoxical as it may seem, even sound businesses may fall into difficulties by too great success; and Carpenter's business was far from sound. At the beginning, it was possible to "wash one hand with the other"; for the profits from the usury branch sufficed to pay the extravagant interest promised to the depositors. But Carpenter know nothing of moderation. He opened a series of branches, one in Bishopsgate, others in the Provinces; and from each of them flowed a stream of cash carrying with it a corresponding liability for interest. What was to be done? Carpenter hit on the simplest solution: pay the interest out of capital and use the residual funds in speculation. Thus if you deposited £100 with Carpenter, he could set aside £7 of it to pay you a first year's interest and utilise the remaining £93 in wild-cat schemes which took his fancy. If his speculations were successful, his business was solvent by his criteria. If the speculations failed. . . . but Carpenter never envisaged failure, since he had a wholehearted belief in his financial acuteness.

His first "investments"—for that was how he regarded them—were in South African shares; and in this field he proved not unlucky. Later came a grandiose scheme for developing the Gaspé Peninsula in New Brunswick. He had learned, it appears, that the harbour at Gaspé was the best on the coast and remained ice-free for two or three months longer than the port of Quebec; and he conceived the plan of converting Gaspé into a great Atlantic terminus. This involved lumber work, saw mills, and the building of a railway to connect Gaspé with the big Canadian lines; and the necessary capital he took from the deposits in The Charing Cross Bank.

It is not impossible that the development of Gaspé might have been a remunerative scheme if it had been properly financed. But on the face of it there would be a long period before expenditure on Gaspé could yield any returns, whilst money would be flowing out in a steady stream for years; and to lock up the money of The Charing Cross Bank depositors in any such long-term enterprise was pure folly, since Carpenter's clients could demand the return of their deposits at any time. But this serves to illustrate the mentality of the projector.

Curiously enough, it was Carpenter himself who precipitated his own downfall. He suspected that employees of his Bishopsgate branch were robbing him, and in 1909 he called in a firm of accountants to sift this matter. They reported that all was correct at Bishopsgate; and Carpenter was so satisfied with this that he asked one of the partners in the firm, Mr. Roy Pembridge, to audit the accounts of the other branches of The Charing Cross Bank. This move is enough to prove two points: first, that Carpenter himself had no glimmer of the fact that his Bank was fundamentally unsound; and, second, that he was, according to his lights, an honest creature, for the last thing a rogue would do would be to reveal to an expert the chaotic condition of his business.

Mr. Pembridge examined the books of several branches of the Bank and discovered that not one of them was making a real profit. He pointed this out to Carpenter, but was apparently brushed aside with the assurance that all was well, as Carpenter had other affairs in hand and was making plenty of money. Mr. Pembridge then began an examination of the state of affairs at the head office in Bedford Street, where he found the bookkeeping in a most unsatisfactory state. He suggested various precautions; and Carpenter, much delighted with the novel idea of running his business in a businesslike way, requested Mr. Pembridge to go over to Canada and examine that side of the concern also. Mr. Pembridge did so, and on his return he informed Carpenter that disaster loomed over the Gaspé enterprise unless certain matters were given immediate attention. Carpenter reacted characteristically by losing his temper; but in the end he had to admit that his Bank was insolvent. He contended, however, that it would be saved from collapse by the profits which the Canadian speculation would eventually bring in, profits which he estimated at the liberal figure of two or three millions. Mr. Pembridge countered by pointing out that in the advertisements asking for deposits Carpenter was mentioning assets of very doubtful value.

But at this stage the auditors found themselves in an awkward position. They knew the instability of The Charing Cross Bank, and they felt that they had a public duty in the matter. They requested Carpenter to put the true position before his clients; but this he refused to do. "It will mean ruin for me, and I shall certainly lose the profit I am going to make." The auditors stuck to their guns: they consulted their solicitors, counsel's opinion was taken, the Director of Public Prosecutions was approached. At last Carpenter was informed that he must either file his petition in bankruptcy or face a criminal prosecution for obtaining money on false pretences. After a few days' consideration, he bade farewell to his dreams and filed his petition.

The investigations of the Official Receiver disclosed liabilities of £2,830,791, with assets—some of doubtful value—totaling £1,131,557. The papers were sent to the Director of Public Prosecutions, and Carpenter came into the dock.

My own impression is that, according to his lights, Carpenter was quite honest in his intentions. He believed outright in his Gaspé scheme, and had not the slightest intention of swindling his depositors. Nor could he be accused of squandering money in luxury, for he lived very quietly at Staines, making no display and esteemed by his neighbours. It is one of those cases in which it is hard to draw a line between dangerous optimism and actual dishonesty. In principle, it differed but little from the case of a bank manager who embezzles money for gambling purposes, hoping to win enough to replace the sum which he has removed from the till. In the end, something like one and three-quarter million pounds vanished in The Charing Cross Bank; many of the depositors were ruined; and in 1911, Carpenter himself, at the age of seventy, was sentenced to two years' hard labour.

Even this drastic lesson failed to awaken the patrons of private banks to the risks they ran. The financial tragedy was repeated on a far larger scale by Farrow's Bank, the downfall of which involved 35,000 clients and four-and-a-half millions sterling. Farrow seems to have been much the same type of man as Carpenter; his methods were similar; but he had a much shorter run, for his Bank was founded in 1907, got into difficulties by 1915, and collapsed in 1920.

Passing across the line which divides dangerous optimism from downright dishonesty, one still finds that in essentials there is a very simple idea at the root of gigantic schemes. In the legend of Perseus, one reads of the three Grey Sisters who possessed jointly a single tooth which they handed about among themselves as occasion demanded. A very similar idea underlay the operations of Whitaker Wright.

Whitaker Wright began as an assayer, and emigrated to the United States at the age of 21, to practise his trade. After an adventurous career in Idaho and elsewhere, he discarded chemistry, went to Philadelphia, and launched into his real life work of company promoting. Two of his ventures—the Sierra Grande Silver Mine and the Colorado Coal and Iron Company—were very successful; he became well known in American financial circles; and at the age of 31, he seems to have become a millionaire. But even at that early stage, clouds gathered about him and he found it advisable to return to England in 1889. Here he turned to the formation of "promoting companies," businesses in which the money of the shareholders is utilised in acquiring properties which can be "floated" instead of being developed. In 1894 he promoted the West Australian Exploration and Finance Corporation, which was fairly successful; and a year later the London and Globe Finance Corporation was formed. A third creation was the Standard Exploration

Company, and there was also the British-American Corporation. From these parent stems branched out numerous "operating" companies upon which was devolved the task of actually developing the properties acquired.

Some of these subsidiary companies were apparently, at least, perfectly sound. The Lake View and Ivanhoe mines in Western Australia paid millions in dividends, and Lake View Consols stood at one time at £23. The heart of the whole affair, however, was diseased.

The details of the transactions were so complex that few laymen could follow them, and the Official Receiver spent twelve months in investigating the affairs of the London and Globe before he could issue even his first report; but the basic idea was simple enough. The central companies were kept afloat by handing their assets to each other in turn, just as the Graiæ passed their tooth from one sister to another. Thus, at the general meeting of Company A, it was found that parts of its assets were represented by claims on Companies B and C. When the time came for a general meeting of Company B, these claims had been juggled with in such a way that now Company B had holdings in Companies A and C; and when, in due course, the affairs of Company C were laid before its shareholders, it was represented as solvent because of its holdings in Companies A and B. Much depended on the skill with which the date of the general meeting was chosen. In one case, had the meeting been held on September 30 instead of on December 5, the balance sheet would have shown the gigantic deficiency of £1,600,000 in the London and Globe accounts; but in the interval between the two dates, Whitaker Wright so juggled with the affairs that he was able to exhibit an apparent profit of £463,372.

It might have been supposed that such manoeuvres could hardly escape the notice of Whitaker Wright's co-Directors; but in choosing these he had shown a skill equal to that which he used in selecting the dates of his general meetings. They bore names well known to the man in the street from public service; among them were an ex-Viceroy of India and an ex-Governor of Victoria; but in company finance they were the merest children, as the sequel showed. Thus, Whitaker Wright secured colleagues whose probity reassured the investor without hampering himself with anyone likely to know a hawk from a handsaw in financial affairs.

Unlike Carpenter, who lived modestly, Whitaker Wright had a passion for display, which found its focus at Lea Park, near Godalming. Fonthill Abbey cost Beckford £273,000, but it was completely eclipsed by the gigantic "folly" on which Whitaker Wright squandered his money. In 1896, he paid a quarter of a million for the estate, and within three years the new owner lavished a million sterling upon it in order to alter it to his liking, and to convert it into what was said to be the finest and most expensive private residence in the world. To enclose the estate, a wall 14 miles in length was built. Italian sculptors were brought to England to execute the statuary and marble fountains which studded the domain. Artificial chasms, grottos and caverns were excavated. At a word from Whitaker Wright a hill vanished from the landscape or a lake appeared. The very soil of the estate was moulded into a long succession of terraces which rose, tier above tier, to the summit of rising ground on which the mansion stood. A private theatre was erected at a cost of £15,000. And, strangest fantasy of all, there was a sublacustrine conservatory with a glass roof constructed under the surface of an artificial lake, lest this modern Aladdin should be incommoded by the heat of summer days.

In 1899, the Lake View mine was producing 30,000 ounces of gold per month, and, according to local information, this output could be kept up indefinitely. On the strength of this information, Whitaker Wright determined to get control of all the Lake View shares. Now at that moment, all the spare cash of the London and Globe Finance Corporation was locked up in the operations of driving the tunnels of the Bakerloo Tube; and in order to buy Lake View shares, then standing at £23, Whitaker Wright had to borrow heavily from his allied companies. Suddenly, when he was involved to the neck, the output of the Lake View mine fell to 10,000 ounces per month, and the shares for which he had paid

£23 dropped to less than £10. Between them, by this unanticipated disaster, the London & Globe and the Standard Exploration Company lost £1,000,000.

Then followed frantic efforts to bolster up the finances of the London & Globe temporarily, in the hope that the Lake View mine would regain its earlier output; and Whitaker Wright resorted to the expedient of making each of the three principal companies lend money to the others as and when required to produce satisfactory balance sheets at general meetings. On a single transaction in Lake View shares, the London & Globe had incurred a loss of £782,000, yet on December 15, a balance sheet was issued purporting to show the Corporation in a flourishing state. Thirteen days later the London & Globe suspended payment.

The crash resounded across the world, for the wreck was complete. The Corporation was capitalised at £2,000,000; but only a few thousand pounds could be produced, and these were used to pay a small dividend to clamorous creditors. The shareholders got nothing. Nor were they the only sufferers. One firm of stockbrokers alone lost £365,000, and was hammered on the Exchange.

The finances of the various companies were in such a complicated state that it was only in 1901 that an order was made to wind up all Whitaker Wright's concerns, and it was not until early in 1903 that the financier realised that the game was up. He fled to Paris; then, accompanied by a young Frenchwoman to amuse his leisure, he sailed for the United States. He was extradited, brought back to London, and put on trial. In January, 1904, after a hearing lasting twelve days, Whitaker Wright was sentenced to seven years' penal servitude. Justice, very slow indeed, but sure, had come upon him. But there was one scene more to play before the curtain fell upon that frenzied career. Whitaker Wright had foreseen the result and had come into Court doubly prepared, with cyanide of potassium in his possession and a loaded revolver in his pocket. He managed to swallow the poison and died in the precincts of the Court while awaiting his transfer to prison.

Just as Carpenter's career found its parallel in Farrow's operations, so Whitaker Wright had a match in Gerald Lee Bevan, who came to grief in 1922. Like Whitaker Wright, Bevan provided himself with some directors more celebrated for probity than for business acumen. In this case also there was an almost inextricable entanglement of the finances of three different companies. In both cases, false balance sheets were used to delude the public, and a firm of stockbrokers suffered heavily when the crash came. Both financiers sought refuge in flight; and in each case the companion in exile was a young Frenchwoman. And both men received the same sentence: seven years' penal servitude.

As a last illustration in this field, the case of Clarence Hatry and his associates may be mentioned, since it exhibited an effrontery in fraud which far surpasses anything described above.

The keystone of the Hatry group of companies was Austin Friars Trust, formed in May, 1927, with a capital of £300,000. After a life of 28 months, the Trust collapsed and was compulsorily wound up in September, 1929, with liabilities totaling some £19,000,000 against assets worth, nominally, £5,500,000. These assets consisted largely of shares in companies associated with Austin Friars Trust; and as these concerns came to grief in the Hatry crash, shares in them were hardly realisable at face value. Apparently they were mere feeders for Austin Friars Trust, which secured the money raised through their agency. Only one of them—Corporation & General Securities, Ltd.—need be mentioned here, as it played a notable part in the disaster.

The tale of the final catastrophe is simple and dramatic. In June, 1929, Hatry and his associates promoted a company called Steel Industries of Great Britain. This was a holding company, and initially it was intended to acquire the shares in two large existent steel companies. The money for this purpose was to be raised from syndicates, and also by borrowing from banks on the security of certain stock. In these ways, about £4,800,000 was obtained; but about £1,500,000 of this sum was spent in meeting liabilities previously incurred by Austin Friars Trust and in buying shares in some of the associated companies. This was the main cause of the trouble which immediately followed; for a frantic effort was necessary to fill the gap left by this £1,500,000.

On Sunday, June 23, 1929, Hatry invited his fellow-directors to his house. The time was close at hand when the holders of the shares in the steel enterprises would have to receive payment for their securities from Steel Industries of Great Britain; and some millions of pounds had to be found to meet these claims. Hatry had bad news for his associates. Certain negotiations had been afoot to raise the necessary capital, and these negotiations were not going well. Harry's view was that the situation could be saved, but it would take time to find the necessary funds. The problem was debated by this group of over-worked, harassed, and excited people, seeking some method of tiding over the danger period.

Among the directors was an Italian, Gialdini, and he threw out a suggestion which changed the course of events. Why not forge a duplicate set of scrip, and by pledging both the real documents and the forgeries, raise the funds necessary to cope with the emergency? With the Latin taste for melodramatic gesture, Gialdini backed his proposal with a threat to blow out his brains before leaving the house, if his plan were not adopted. Hatry, at his wits' end, fell in with Gialdini's views; the protests of the other directors were brushed aside; and what had been, up to that moment a mere question of unlucky company business, was changed into a criminal conspiracy.

One example of the method employed in practice will suffice. The City of Swindon had issued £500,000 worth of Corporation Stock. One of the Hatry companies—Corporation & General Securities—arranged to buy this block at a discount, and handed to the Swindon authorities a first cheque for £250,000. This cheque was all that Swindon ever obtained in the transaction. With the Swindon scrip in their hands, Hatry and his associates duplicated it; and then proceeded to raise money on both the genuine and the spurious documents. By pledging the genuine scrip, they borrowed £450,000, whilst on the bogus scrip they obtained £220,000. Thus, by giving a cheque for £250,000 to the Swindon authorities, Hatry and his friends had been able to raise £672,000, leaving a balance in their hands of £422,000. Ultimately, the losers were those firms which advanced the £220,000 on the bogus security.

Time was, of course, the essence of the whole scheme. Ere long, the holders of the forged scrip were sure to send it in for registration, and then the cat would be out of the bag. But the conspirators hoped that before then they might have carried their negotiations for funds through to success and that they could recover the forged documents before trouble arose. They resorted to various methods for postponing the evil day of registration; but time failed them.

On 18th September, 1929, various banks instructed the well-known chartered accountant, Sir Gilbert Garnsey, to investigate the affairs of the Hatry group of companies. Hatry must have realised instantly that the game was up. By telephone, he made an appointment with Sir Gilbert for the following day; and at this meeting he and three other directors frankly volunteered that they had been guilty of serious offences. On Sir Gilbert's advice, they presented themselves on the following morning at the office of the Director of Public Prosecutions, where they were arrested. Their trial began before Mr. Justice Avory on 20th January, 1930. Curiously enough, they pleaded "Not Guilty"; but on the fourth day they withdrew this and pleaded "Guilty," thus bringing the proceedings to an abrupt close.

One point came out during the trial. None of the defendants had taken a single penny of the money they had fraudulently raised. In fact, Hatry had contributed every cent in his possession to keep things afloat and had made himself a bankrupt as a result. A casuist would find much of interest to him in this curious case of "honest" dishonesty, whilst the cynic might smile to see a set of men trafficking unscrupulously in millions without bettering themselves by so much as a shilling in the process.

What the Judge thought of it was plain enough. He was "unable to imagine a worse case" than that of Hatry, whom he sentenced to the maximum 14 years' penal servitude under the statute. The other three directors were awarded seven, five, and three years' penal servitude respectively. An appeal was made, but it was unsuccessful.

Of the five directors of Austin Friars Trust, one was not present in the dock with the others: Gialdini. After the melodramatic side of his Latin character had been exhibited at the meeting in Hatry's house, the other characteristic of the Latin—prudence—seems to have got the upper hand. On the collapse of the conspiracy, he did not blow his brains out as he had once threatened to do. Instead, prosaically, he took a ticket to his native country, Italy, whence he could not be extradited to stand trial along with his colleagues. Some people—the Attorney-General amongst them—regretted his absence from the scene.

VI
THE WONDERFUL LAMP

MY RECOLLECTIONS OF READING go back to those golden days when, in my father's study, I struggled, text in hand, to convey to him the enthralling news that "The Cat sat on the Mat, and the Rat came in." He was very patient. He even feigned an interest which he must have been far from feeling. But he was always like that: understanding, sympathetic, and ever ready to smooth away difficulties.

A year or two later, I discovered that the books which lined the walls were not without their interest, even to so juvenile an animal as myself. *The Pilgrim's Progress* rubbed shoulders with Jules Verne. There were illustrated editions of Mark Twain in which the pictures stimulated the young reader's curiosity and kept him reading. There was a *Joe Miller's Jest-book*, probably bought to amuse my elder brothers. Encyclopaedias, astronomical maps, and works on Natural History opened a field for a young mind in search of hard facts about everything. There was a Brewer's *Reader's Handbook* which familiarised me with the plots of novels which I shall never read; and which, later, served as a guide in my demands for books from the University Library.

There were some battered, paper-covered volumes of the early Kipling in the old Wheeler Indian edition which gradually got read to tatters and vanished into the waste-paper basket, so little did one foresee their future value. Sometimes I regret that I came to Kipling so young. To a child not yet in its 'teens, Mrs. Hauksbee looked sere if not venerable; and I have never managed to reconstitute the impression left on me by these juvenescent perusals.

As for Mrs. Herriott in *The Story of the Gadsbys*, I could make neither head nor tail of what the woman was talking about. The Age of Innocence has its drawbacks.

That library had other virtues. My father edited the University Calendar, and a serried row of its grey-bound volumes stretched along one shelf. Their contents had no interest for me; a glance or two satisfied me on that point. But they had other uses. The springy backs made them admirable materials from which to build up the cushions of an improvised nursery billiard-table on which one could bring off breaks of eight or ten with walking-sticks for cues and three old "gutty" balls. I doubt if anyone, except myself and my playmates, got much amusement out of these severely practical volumes. Which shows that even the dullest books may yield enjoyment if properly utilised.

But of all the volumes in that library, the one which I remember best was an old, battered, green-bound copy of *The Arabian Nights*, on which I stumbled one day and devoured with delight. It was dingy; it was closely-printed without an enlivening illustration; and the text flaunted its Gallic origin in many a curious twist of idiom which lent it a faintly exotic flavour. But who would stop to consider such trifles when a new world was opening before one's eyes? One could share in the night-wanderings of the Sultan in Old Baghdad, or voyage with Sindbad, or listen to the interminable chatter of the loquacious Barber, or shiver with Ali Baba when he found his brother's body, or mount the Ebony Horse and ride the skies long before the coming of the aeroplanes, or enter Aladdin's cave of jewels in search of the Wonderful Lamp: all at the cost of straining one's eyes a little over the small type on the pages. But the names were terrible. Scheherazade! Golly! what a mouthful and what a tangler for a juvenile tongue! It was with no little relief that I learned, in later years, that it was more accurate—and much easier—to call the lady Shahrazad. So Shahrazad she remains, so far as I am concerned. For this relief, much thanks.

A complete edition of *The Arabian Nights* contains over 250 tales. Where did these originate? When were they written? And who invented them?

As to the place of origin, there seems little doubt. The "Herodotus of the Arabs," Al-Mas'ûdi, writing in A.D. 944, refers to "books which have come down to us" (the Arabs) translated from Persian; and among these he specifically mentions "the book entitled *Hazar Afsanah or the Thousand Tales*, which is known to the public under the name of *The Book of a Thousand Nights and a Night* (*Kitab Alf Laylah wa Laylah*)." Further, in A.D. 987 was published a Catalogue of Arabic Works (*Kitab-al-Fihrist*), which describes the nature of *Hazar Afsanah* sufficiently to make it plain that its framework was the same as that of *The Arabian Nights*. According to other confirmatory evidence, the *Hazar Afsanah* was composed by (or by order of) the Persian Queen Humai, who was thus a forerunner of Queen Marguerite of Navarre with her *Heptameron*. Unfortunately, the *Hazar Afsanah* seems to have vanished like the Milesian Tales of Antonius Diogenes, leaving only scraps behind, embedded in the texts of other authors.

But the *Hazar Afsanah* furnished only the groundwork of *The Arabian Nights*, which grew by degrees to its present form. Some additions were apparently made in the eighth century; further accretions date from the tenth century; and the most modern contributions were made as late as the sixteenth century.

How can we infer this? Take a case in illustration. If *Vanity Fair* were put into the hands of a schoolboy and he were asked the earliest date at which it could have been written, he would have no difficulty in saying: "1815, since it mentions the Battle of Waterloo, and could not have been written before that." If a diary mentions a visit to the Albert Hall, the entry must have been written after 1870, since there was no Albert Hall until 1871. On the other hand, if a diarist describes a journey by stage-coach, it is fair evidence that he wrote before railways displaced the coach as a means of transit. If he speaks of the doctor arriving in his brougham, it is easy to infer that his diary was written before, say, 1910. Everyone remembers how often the hansom cab appears in *Sherlock Holmes;* and where is the hansom now?

The Arabian Nights, then, had no single author. With its germ in the Persian *Hazar Afsanah*, it passed over into Arabic, and came down the centuries, picking up an addition here, adding an accretion there, until it took the form which we now know. But how did these accessions come to be tacked on to the original material? Here one encounters an idiosyncrasy of the Asiatic: he loves to be told a tale. One sees this in *The Arabian Nights* itself, where the Sultan has to be soothed to sleep with a story; but the practice goes far further back than that. "On that night could not the king sleep, and he commanded to bring the book of records of the chronicles; and they were read before the king." So one learns from the Book of Esther, which shows that practice is of very respectable antiquity.

Kings and Sultans may order as they please, but how did the general public gratify its penchant for listening to stories? The answer to that is found in the existence of the Ráwi or professional narrator of such things. Dr. Scott describes him thus: "He recites walking to and fro in the middle of the coffee-room, stopping only now and then, when the expression requires some emphatical attitude. He is commonly heard with great attention; and not infrequently in the midst of some interesting adventure, when the expectation of the audience is raised to the highest pitch, he breaks off abruptly and makes his escape, leaving both his hero or heroine and his audience in the utmost embarrassment. . . . The auditors suspending their curiosity are induced to return at the same time next day to hear the sequel."

Some of these story-tellers have marked dramatic ability. Sir Richard Burton (to whose works I am indebted for many of the details used here) has described one whom he saw in Tangier. "(He) opens the drama with extempore prayer, proving that he and the audience are good Moslems: he speaks slowly and with emphasis, varying the diction with breaks of animation, abundant action, and the most comical grimace: he advances, retires, wheels about, illustrating every point with pantomime; and his features and gestures are so expressive that even Europeans who cannot understand a word of Arabic divine the meaning of his tale. . . . The performance usually ends with the embryo actor going round for alms and flourishing in the air every silver bit, the usual honorarium being a few 'flits,' that marvellous money of Barbary, big coppers worth one-twelfth of a penny."

"In considering the style of The Nights," says Sir Richard, "we must bear in mind that the work has never been edited according to our ideas of the process. . . . (But) even the vagaries of editor and scribe will not account for all the incoherences, disorder, and inconsequence, and for the vain iterations which suggest that the author has forgotten what he said. . . . The characteristics of the whole are naiveté and simplicity, clearness, and a singular concision. The gorgeousness is in the imagery, not in the language; the words are weak, while the sense, as in classical Scandinavian books, is strong. . . . I cannot take up the Nights, in their present condition, without feeling that the work has been written down from the Ráwi, the conteur or professional story-teller."

When one considers the manner in which the collection has been brought together, it is not surprising to find that various manuscript versions of *The Arabian Nights* differ markedly from each other not only in arrangement but even in subject-matter, especially since these versions became rarer and less accessible to the Ráwis. Lane published his *Manners and Customs of the Modern Egyptians* in 1836, and he notes "the great scarcity" of copies, "even fragments of them are with difficulty procured; and when a complete copy of 'The Thousand and One Nights' is found, the price demanded for it is far too great for a reciter to have it in his power to pay." From this modern instance, one may guess how, in earlier times, invention had to step in and add to the congeries when the Ráwi fell short of material for his recitations. And one can understand also how the language of the book has become a most curious hotch-potch of the ornate, the mean colloquialism, and the euphuistic.

How did *The Arabian Nights* come into Europe? Here, fortunately, there is no dubiety. The honour of enriching thousands of Western nurseries, and of charming elder intelligences to boot, belongs incontestably to Antoine Galland. His own life was something of a romance. He was born in 1646 of peasant parents at

Rollot, a couple of leagues from Montdidier. His mother, widowed in early life and compelled to earn her own living, had no means of educating young Antoine, who was her seventh child. Fortunately, the President of the Collège de Noyon stepped in; and for nine or ten years Galland studied in the Collège, where he learned Latin, Greek, and even Hebrew by the age of fourteen. Then, through no fault of his own, his schooling ceased and he was apprenticed to a trade. Had he lacked character, he would probably have ended up as a handicraftsman; but he was cut out for better things. After a year spent in vexation of spirit, he escaped to Paris; and in those days it was no light matter for an apprentice to shake free. His only acquaintance in the capital was an aged kinswoman. She bestirred herself on his behalf, and he was received in the Collège du Plessis, where his progress in Oriental studies gained him a transfer to the Sorbonne. Here again he showed such marked ability that he was appointed to catalogue the Eastern manuscripts in the possession of that institution. After a short tenure of a post in the Collège Mazarin, he was appointed attaché-secretary to the Marquis de Nointel, diplomat and antiquary; and in 1670 Galland went to the Levant with his chief, who had been nominated French Ambassador at Constantinople. So the peasant's son came to that East which was to spread his name all over Europe.

Galland's duty was to study dogmas and doctrines, always knotty subjects in that region, as the Crimean War serves to remind us. He frequented the cafés, mastered the language, listened to the story-tellers, and grappled with religious topics. M. de Nointel, having settled some diplomatic matters, betook himself to the "Holy Places" of the Levant, including Jerusalem; and Galland followed in his train, copying inscriptions, sketching monuments, and collecting antiques. Meanwhile, by his study of Eastern men and manners, he was preparing himself for the work which was to render him renowned. In all, he made three journeys to the Levant, and gained a thorough knowledge of the Arabic, Persian, and Turkish languages and literatures. He was a lucky man; for in Smyrna he had a narrow escape during the earthquake and fire which cost the lives of 15,000 people. Galland himself was buried in the ruins, "but his kitchen being cold, as becomes a philosopher's, he was dug out unburnt."

Returning to France, where he settled down in Caen, he issued in 1704 six volumes of his *Mille et une Nuit, Contes arabes traduits en francais*. And, straightway, this unobtrusive scholar flashed into celebrity.

How did this come about? The French are conservative in many ways, and especially in the matter of foreign literary works. *The Thousand Nights and a Night* was literature of a kind which they had never before encountered. Why, then, did it score an instantaneous success with the French reading public? The answer to this lies in Galland's character. Had the book been presented to the French reader by a pedant, intent only on producing an accurate literal translation, the chances are that it would have fallen flat on publication and become a drug in the market instead of a universal success. But Galland was no pedant, despite his learning; and by amazing good fortune he was a born teller of tales. Filled with enthusiasm for his subject, he aimed at interesting the public in this gift from the East. The precisian may hold up his hands in horror at "the sins of omission and commission, of abridgement, amplification, and substitution, and the audacious distortion of fact and phrase in which Galland freely indulged"; but common sense looks at Galland's achievement and finds it good, whatever minor defects it may have had from the standpoint of the purist. Concede to Galland the rights which the Ráwi claimed in reproducing the stories for an audience, and no fault can be found with the method of presentation which he adopted.

Nor was he slow to take a hint, even during the publication of the work. In the first two volumes, he preserved the mechanism of the Arabic, so that at the beginning of each night appeared the invocation of Dunyazad to Shahrazad: "My sister dear, if you be not sleepy, pray tell us one of those stories which you know so well." Some young folk, a-weary of these repetitions, resolved to make a practical protest. Accordingly, one chilly night, they knocked at Galland's door and disturbed the slumbers of the distinguished author. Clothed only in his night-shirt, he hurried to his window to find what the noise was about; and having got him there, the young people kept him in the cold by asking one trivial question after another. At last, having exasperated him, they ended the comedy by saying: "Ah, Monsieur Galland, if you be not sleepy, pray tell us one of those stories which you know so well." This veiled protest seems to have had its effect; for in the subsequent volumes the exordium was dropped and, instead, the text was broken by inserting the appropriate number of the night.

In 1706, Galland, now a well-known man, returned to Paris to pursue his researches. Three years later he became Professor of Arabic in the Collège de France. He was, in real life, the embodiment of the popular idea of a professor. Scrupulous in all things, single-hearted in his devotion to letters, he combined penetration in his studies with a complete simplicity in his life and manners, and a certain childishness in the affairs of the world. He would have been as happy in teaching children as he was in acquiring his immense erudition.

During his lifetime he published only six out of a projected twelve volumes of *The Arabian Nights*. In his last illness, the ruling passion was still strong, for he sent to Noyon for his nephew to help him in arranging his manuscripts and making his will. He died on February 17, 1715. The final part of his edition of the *Nights* appeared two years later, and in it was contained a series of the best-known tales: *Aladdin, Ali Baba and the Forty Thieves, Prince Ahmed and the Fairy Peri-Banu*, as well as others. We shall see later some of the problems to which these gave rise.

Translations of Galland's work soon spread *The Arabian Nights* beyond the frontiers of France. Thus, in *The Spectator* of 1712, Addison embodied the story of Alnaschar, describing it as "an Arabian tale which I find translated into French by Monsieur Galland." In 1720, Pope sent a couple of volumes to Bishop Atterbury, who seems not to have been pleased. The fascination which the work exercised on some minds can be guessed from an anecdote of Sir James Stewart, Lord Advocate of Scotland. One Saturday evening he discovered his daughters deep in *The Arabian Nights*. Those were strict times, and the Lord Advocate was properly shocked. He confiscated the volumes and bade his daughters remember that the evening before the Sabbath should not be devoted to any such worldly matters. But alas for virtue! Sir James opened the books, fell under their spell, and was found still engrossed in them on the morning of the Sabbath itself, for he had been unable to lay them down and had spent the whole night reading them. Another Scottish intellect was less receptive. Thomas Carlyle described *The Arabian Nights* as "downright lies," and refused to have such "unwholesome literature" in his house. This, from the author of *Frederick the Great*, is illuminating in more ways than one.

But we must hie back to France and continue the tale of *The Arabian Nights* themselves. There came to Paris a certain Al-Kahin Diyûnisiûs Sháwísh, a Syrian priest of the Congregation of St. Basil. He became a teacher of Arabic, and his somewhat rebarbative cognomen was smoothed down, so that he became known as Dom Chavis. He had brought with him from the East an Arabic manuscript cognate to *The Arabian Nights*, and he set about translating certain Oriental tales. Unfortunately, he seems to have been something of a polyglot, and unable to keep one language clear of another, so that his efforts resulted in a queer Arabo-Franco-Italian hybrid which was hardly likely to be of popular interest.

At this juncture, Dom Chavis forgathered with Jacques Cazotte, who consented to help the Syrian by converting his linguistic monster into comprehensible French.* Cazotte himself was a curious character. Born at Dijon in 1720, he was educated by the Jesuits, and went into the public service. At twenty-seven he became Comptroller of the Windward Islands, where he showed considerable courage when the British attacked St. Pierre. He acquired estates and slaves in the Antilles and married the daughter of the Chief Justice of Martinique.

---

* One is reminded of Butcher and Lang's translation of the *Odyssey* which was wickedly described as "translated from the Greek by Butcher and done into English by Lang."

Meanwhile he had tried his hand at literature; and had made a certain reputation in both prose and verse, composing tales, songs, and even an opera—*The Thousand and One Fopperies*, which showed the influence of *The Arabian Nights*. During his residence abroad, he wrote a mock-heroic poem in twelve cantos, the *Roman d'Olivier*, which had marked success, and this was followed by other works which gained a wide circulation. He was clever with his pen, for among his writings was a continuation (or parody) of Voltaire's *Guerre civile de Genève*, composed with such verisimilitude that it deceived the public completely into imagining that Voltaire was its author.

Sickness drove him from his post in Martinique, and he returned to France to learn, on landing, that his brother had died, leaving him an estate near Epernay. This piece of unexpected good fortune was, however, balanced by a financial disaster; for the friend whom he had entrusted with the care of his estates in the Antilles proved unfortunate in speculation, and Cazotte thus lost the whole of his savings, some fifty thousand crowns.

In Paris, Cazotte was caught up in the whirl of gaiety, and his kindheartedness, his wit, and his social qualities made him a favourite. Later in life, though still keeping touch with the society of the time, Cazotte became an ardent Christian, by no means an easy role in the fashionable Paris of that day. He is reported to have embraced the tenets of the Illuminati and to have declared that he possessed the power of prophecy.

Politically, he was a Royalist, and this led to his final disaster. The French Revolution was looming over the land, and Cazotte was one of its victims. By nature he was frankness itself, and in letters to his old friend Ponteau, Secretary of the Civil List, he made no secret of his opinions. In August, 1792, these letters unfortunately came to light; and the old man—for he was now over seventy—was thrown into prison. On September 3 he was attacked by patriot-assassins, and though he escaped for a time, through the devotion and courage of his daughter, he was once more arrested and sent before a revolutionary tribunal. During an interrogatory of thirty-six hours, his serenity and presence of mind impressed even his accusers. But "it is not enough to be a good son, a good husband, and a good father; one must also prove oneself a good citizen." He was condemned to death. He spent his last hours with his confessor, to whom he entrusted, along with a lock of his hair for his beloved daughter, letters to his wife and children, in which he begged them not to mourn him, not to forget him, and never to offend against their God. On the scaffold, his last words were: "I die as I have lived, faithful to my God and my King."

This was the man who collaborated with Dom Chavis in turning the new Arabic manuscript into French. Though he was then nearing seventy, Cazotte had in his literary affairs all the vigour of youth; and after midnight, returning from his social pleasures, he took up his pen and worked on until the dawn. He chose Galland as his model in working up his material to suit the style and taste of his day; but he went far further than his prototype, for instead of confining himself to the material provided by Dom Chavis, he let loose his unwearying powers of invention and interpolated tales entirely his own, with the object of spreading his spiritual ideas under the Arabian guise. So far did he go, indeed, that only the existence of the Dom Chavis original Arabic manuscript preserved in Paris serves us to distinguish between his translation and his invention.

At this period fairy tales were in vogue; and a great collection of these, the *Cabinet des Fées*, was being published in Geneva, between 1788 and 1793, comprising forty-one volumes in all. Of these, the last four volumes contained the work of Chavis and Cazotte, described as *A Continuation of the Thousand and One Nights*.

We may now turn to the history of *The Arabian Nights* in English dress. During the two centuries and more which have elapsed since Galland published his French translation, various English renderings of it have come steadily from the press and are still coming: a practical proof that Galland's work fills the requirements of a very large public, for few authors can claim such a circulation after so long a span of time.

There is, however, another translation from a French version which deserves notice here. In 1899, Dr. J. C. Mardrus began the publication of his French translation from the Arabic, which aimed at a literal rendering of the original. This Mardrus translation has been done into English by the late Edward Powys Mathers, and his *Book of the Thousand Nights and One Night* appeared first in sixteen volumes (1923-4) and then in a revised four-volume guise in 1937. Mathers was specially fitted for his task as translator. He had a light touch in the versified portions of the work—probably the most difficult to reproduce without wearying the reader—and he completely avoided the Gallicism which gives a foreign twang to the usual English translations of Galland His object was to produce a simple unannotated version of the complete work for the entertainment of the casual reader; and in this he was entirely successful, for his rendering is the most readable yet produced for those who seek a clear narrative in good and fluent English. His version of *The Arabian Nights* was intended for adults, it may be said, and not for the schoolroom or the nursery. Mathers was something of a polymath. In the course of a too-short life, he turned his hand in succession to poetry, prose, play-writing, and translation from foreign tongues, whilst, under his pseudonym "Torquemada," he won the gratitude of lovers of crossword puzzles.

Excellent though it may be, however, any translation of a translation is inevitably like the echo of an echo; and we must now turn to those English versions taken directly from the original Arabic.

The pioneer of direct translation from Arabic into English was Dr. Jonathan Scott, who, in 1811, brought out a version based on the Wortley-Montague manuscript. But Scott, like Cazotte, chose Galland as his model and, in practice, outgallanded Galland in the liberties he took with his text. Then, in 1838, Henry Torrens began a literal translation; but only a single volume was published, though on the same scale the work would have required nine or ten volumes for its completion. Edward William Lane, a grandnephew of Gainsborough, and one of the most distinguished Arabists of his day, produced, in 1839-41, an exact translation, accompanied by a mass of most valuable notes on Moslem manners and customs; but it was far from complete. He took as a basis the abbreviated Bulak edition of the *Nights*, and as his work was avowedly intended for the "drawing-room table," he was forced to omit a good deal of material unsuitable for the early Victorian drawing-room. Out of a possible 267 tales, he translated only 106. The field was thus left open to John Payne, who printed, for private circulation only, five hundred copies of the first complete and literal translation of the great collection, in nine volumes.

Payne gave an undertaking to his subscribers that he would not reproduce his work "in its complete and uncastrated form," so it was practically unprocurable after his limited edition had been distributed to his clients. But, as the event proved, a public existed outside the limits of the favoured five hundred. Who would feed these hungry sheep? Into the breach stepped one of the most flamboyant personalities of the period. From the British Consulate in Trieste came a lithographed circular:—

> "Captain Burton, having neither agent nor publisher for his forthcoming ARABIAN NIGHTS, requests that all subscribers will kindly send their names and addresses to him personally (Captain Burton, Trieste, Austria), when they will be entered into a book kept for the purpose.
> "There will be 10 volumes at a guinea a piece. . . . Captain Burton pledges himself to furnish copies to all subscribers who address themselves to him; and he also undertakes not to issue, nor to allow the issue of a cheaper edition. One thousand copies will be printed. . . ."

What unorthodox person was this, who flouted the normal procedure of the book-trade, thrust publisher and bookseller aside, and presented himself directly to his public in this unceremonious fashion?

Like Beddoes of *Death's Jest-Book*, Richard Francis Burton was born out of his proper time, for by nature he belonged to the Elizabethans. He had their fiery temperament, their fearlessness, their impatience of control, their lust for crossing familiar frontiers and breaking into unknown lands, their love of literature, and other qualities as well. In the sixteenth century he might have left behind him a name but little less illustrious than those of Drake and Raleigh. Cooped up in the Victorian age, he proved himself a misfit; and, after a career of exploration which made him famous, he was left to fret away his life in the obscurity of the Consulate at Trieste.

He was born in 1821. His father, Colonel Burton, suffered from severe bronchial asthma, and was forced to go on half-pay while in prime of life; and thereafter the family roamed the Continent in search of a congenial climate. The young Burton brothers had hardly any formal schooling; but in their travels they picked up languages and dialects as they went along, and they had tutors. Richard learned fencing, and a love of swordsmanship remained with him.

He was an affectionate but unruly boy, and during the family's stay in Sorrento he proved unusually troublesome. He crept over the Natural Arch merely because some Italian lad declared: "Non e possibile, Signorino," and Burton wished to prove him wrong. In those times and much later, it was the fashion for the guide at the Grotta del Cane to prove the existence of a lake of carbon dioxide in the cave by hauling a wretched dog into the heavy gas until it was half-stifled, his own head being well above the surface all the time. Burton insisted on taking the place of the dog, and was pulled out just in time to avoid suffocation. Again, he was detected in an attempt to descend into the crater of Mount Vesuvius. Neighbours had told the boy that Satan had been seen going into the depths, clutching the soul of a usurer; and Burton apparently thought this an excellent chance to make the Devil's acquaintance.

By the time Burton reached nineteen, the two boys had become impossible as companions to their invalid father. They had lavish pocket-money and they spent it lavishly. They walked, they drove, they practised fencing and pistol-shooting, "made experiments in everything imaginable," including opium and love-affairs; but of serious work they did none. "We had thoroughly mastered our tutor, threw our books out of the window if he attempted to give a lesson in Greek or Latin, and applied ourselves with ardour to Pigault-Lebrun and Paul de Kock." "It was evident that the Burton family was ripe for a break-up."

Colonel Burton, casting round for a. profession which would secure a competence, decided to put his sons into the Church. In preparation for this, they were sent to England, and there they were separated. The younger brother, Edward, was placed under the charge of a Rev. Mr. Havergal, rector of a country parish. At first the two boys were allowed to send letters to each other; but this ceased under protest from the rector, who wrote to say that "Richard must not correspond with his brother, as he had turned his (Mr. Havergal's) name into a peculiar form of ridicule."

Meanwhile, Richard had been examined by some professor to discover the scope of his knowledge. "The professor put me through my paces in Virgil and Homer, and found me lamentably deficient. I did not even know who Isis was! Worse still, it was found out that I, who spoke French and Italian and their dialects like a native, who had a considerable smattering of Bearnais, Spanish, and Provençale, barely knew the Lord's Prayer, broke down in the Apostles' Creed, and had never heard of the Thirty-nine Articles." Truly a valuable prospective candidate for Holy Orders!

Colonel Burton, however, stuck to his project. Edward was sent to Cambridge and his elder brother entered Oxford at the age of nineteen and a half. He loathed it, as might be expected of a youngster taken from a life of freedom and thrust into the restrictions which hem in the undergraduate. He was allotted "a couple of dog-holes called rooms." He had cultivated a fine moustache; the authorities issued formal orders that he must shave it off. After Continental cookery, he found the College meals dreadful. The continual clangour of bells irritated his nerves, strong as these were. He pronounced Latin and Greek after the Continental fashion, and this irked his instructors. Wearying of these languages he decided to study Arabic. The University boasted a Professor of Arabic, but he considered it his duty to teach a class but not an individual; and as no one else seems to have hungered for the subject, Burton was left to teach himself.

Besides the recreations of walking, rowing, boxing, and fencing, he amused himself with "an air-cane," which he employed to shoot unhappy rooks over the heads of the dons while they played at bowls. "The grave and reverend signiors would take up the body and gravely debate what had caused the sudden death, when a warm stream of blood trickling into their shirts explained it only too clearly."

Failing to convince his father that he had no aptitude for University life, Burton sought a way to break his fetters, and determined to get rusticated. A steeplechase which undergraduates were forbidden to attend provided his opportunity, and, with some kindred spirits, he cut a compulsory lecture, attended the meeting, and next morning was brought up for judgement. His companions got off with rustication, but Burton was singled out by a special recommendation not to return to Oxford. Doubtless the dons heaved a sigh of relief in thus getting rid of their turbulent pupil; as for Burton, he had achieved his object, and thereafter he refrained entirely from any praise of Oxford.

Edward had proved himself equally intractable at Cambridge. At the end of his first term it was remarked that he was never present at chapel; and his tutor, sending for him, rebuked him in strong language. "My dear Sir," was the unperturbed reply, "no party of pleasure ever gets me out of bed before ten o'clock. Do you *really, really* think that I am going to be in chapel at eight o'clock?" "Are you joking?" demanded the tutor, "or is that your mature decision?" "My very ripest decision," said Edward. He was obliged to leave college without delay.

As soon as Colonel Burton—a very irascible individual—recovered from the news of Burton's planned freak, he bestirred himself on behalf of the scapegrace; and Burton finally began his military career under John Company in the 18th Bombay Native Infantry, but active service was denied him. He arrived in India just too late for the Afghan War; despite his efforts, he was left out of the Scinde campaign; and most of his stay was spent in travel, learning Eastern languages, and studying native customs.

He had a wonderful aptitude for disguise; and, posing as a half-Arab half-Iranian merchant (to avoid difficulties with accent), he was able to mix with the native population without detection. Possibly Kipling's Strickland owed something to memories of Burton. "With hair falling upon his shoulders, a long beard, face and hands, arms and feet, stained with a thin coat of henna, Mirza Abdullah of Bushire—your humble servant—set out upon many and many a trip. He was a *bazzaz*, a vendor of fine linen, calicoes and muslins—such chapmen are sometimes admitted to display their wares, even in the sacred harem, by 'fast' and fashionable dames." In such guise he was sent out amongst the wild tribes of the hills to collect information for Sir Charles Napier.

During one of these wanderings he encountered a beautiful Persian girl of high descent with whom he fell violently in love, and whom he would have married and brought home to England had she lived, for she also was passionately in love with him. But she died, and her loss seems to have stamped Burton for many years.

Another of his adventures in disguise culminated in an attempt to rescue a rather bored young nun from a convent in Goa; but by a misapprehension of the topography of the convent, Burton blundered into the room of the sub-prioress, with results which might have been disastrous. Finding that he had "rescued" something much less attractive than the original quarry, Burton made off and escaped unscathed.

India had not yielded all that Burton expected. His chief gain from his stay was the knowledge he gained of native habits and customs, and a mastery of languages. He passed official examinations in eight of these during his career, and later in life he claimed to know twenty-nine in all, though one may doubt if he really knew all of them perfectly. He certainly broke fresh ground in this field, for he determined to learn the language of monkeys, forty of which he kept as pets. He obtained sixty simian words which he wrote down, but, unfortunately, this list was destroyed in a fire at a later period. Finally, an attack of rheumatic ophthalmia put an end to his usefulness for the time being, and he was given leave to return to England, broken in health, after what he then regarded as seven wasted years.

He had not made himself popular He indulged in eccentricities of dress, roughness of manner, hot temper, a disposition to wage war on harmless prejudices; and, above all, he gave full reign to the "almost suicidal practice" of telling horrible tales against himself, little realising how seriously this form of "leg-pulling" would react against him. He would relate some ghastly story of having eaten a boy, or shot two or three men for no valid reason, or run away with half-a-dozen of other people's wives: and all this rubbish was duly scored up against his reputation as if it were the truth. He grew more and more discontented with his commonplace environment, and cast about for some grand adventure.

Burckhardt was one of his heroes; Burckhardt had penetrated to the forbidden city of Mecca in disguise: Burton would do likewise. But Burton's original plan had a far wider scope; he proposed not only to visit the Moslem Holy Cities but to explore the deserts which lie between Mecca and Muscat. He offered his services to the Royal Geographical Society, who favoured his project, but when they approached his military superiors to ask that Burton be granted the necessary leave, they met a refusal. Apparently the proposed full plan was regarded as too dangerous; and all that was granted was an extra year's furlough "that he might pursue his Arabic studies in the lands where the language is best learned," a cautious phraseology covering a journey to the Holy Places but not implicating the authorities directly.

There is no space here to deal with the details of that journey. It required, in those days of fanaticism, a colossal preparation against mistakes, for even the most trifling blunder might cost an explorer his life. It may seem a simple thing for a European to drink a cup of water; but for an Indian Moslem that act involves no less than five formalities unknown to the West. Multiply this complexity by carrying it into the most ordinary acts of life, and the forestalling of possible errors becomes a giant task.

Burton left England in his old disguise of Mirza Abdullah of Bushire, but in Alexandria he became a physician, the Shaykh

Abdullah, born in India of Afghan parents. Taking camel he crossed the desert to Suez and embarked on an open boat of about fifty tons, carrying ninety-seven pilgrims, all barefooted, bareheaded, dirty, ferocious and armed. There was a scuffle aboard, but Burton reached Yambu, the port of El Medinah, in safety. Here he joined a caravan escorted by irregular cavalry. On the road to Medinah they were attacked by Bedawi, had to fight most of the way, lost twelve men, and reached their destination in eight days. It was no comfortable journey; Burton was lame through the prick of a poisonous thorn. At Medinah he saw the tomb of the Prophet and other spots of pious interest. Then, after a stay of six weeks, he departed for Mecca.

On the road to Mecca the pilgrims suffered from the dreadful heat of the simoom, the deceptions of the mirage, the menace of whirling sand-pillars, and the pangs of thirst. Many of the beasts of burden died and fell to the ever-watchful vultures which hung aloft, ready to pounce. Nor was it a peaceful company. One pilgrim stabbed another in a pointless quarrel and the victim, mortally wounded, was wrapped in his shroud and left to die in a half-dug grave. At El Zaribah, they donned the proper ceremonial garb; but even then their troubles were not over, for they had to beat off another attack delivered by Bedawi, by night, in a grim ravine.

At Mecca, Burton performed all the proper ceremonies: he visited the Ka'abah and paced seven times round it; he touched the sacred Black Stone; he drank the holy but bitter waters of Zemzem; he saw the Praying Place of Abraham, Mount Arafat, the Tomb of Adam; and he stoned the Devil at Muna. On September 26 he embarked at Jeddah on the ship which brought him back to Suez.

In after-life Burton was never ostracised by Moslems, but rather the contrary was the case. Yet for a Christian to penetrate to the Sacred Places in those days would have been sacrilege of the worst. Presumably he was regarded as a genuine follower of the Prophet. Certainly some of his opinions pointed in that direction. But Burton's religious tenets remained a mystery throughout his life. Probably he belonged to what Disraeli called "the religion of all sensible men." ("And what is that?" "Sensible men never tell.")

This perilous expedition made Burton's name famous, and he cast about for further adventure. Another forbidden city beckoned to him: Harar, the capital of Somaliland, which no European had yet entered, despite numerous attempts. John Company had long coveted Berberah and desired accurate information about the country; and they willingly granted Burton a further furlough for this purpose. Three other men were allotted minor parts in the plan: Lieutenants Speke, Herne, and Stroyan; but Burton was given the most difficult and dangerous task, a penetration into Harar itself. He reached the capital, visited the Amir whose least word was death, transacted some political business and, after a short stay during which he gathered much valuable information, Burton started on his return to Berberah. It was a dreadful journey through unfriendly country, and Burton would have done well to wash his hands of East Africa at that stage. Instead he planned a second expedition on a larger scale. It met with total disaster. Hardly had the company landed at Berberah again than a fight broke out: Stroyan was killed, Speke received eleven wounds, Burton had his jaw transfixed by a Somali lance which carried away four back teeth and part of his palate.

As soon as he had recovered from his wound Burton volunteered for service in the Crimean War, where he served under Beatson. He devised a plan for the relief of Kars, and this brought him to the notice of that formidable ambassador, Lord Stratford de Redcliffe. "I felt the game was in my hands, and proceeded to submit my plan for the relief of Kars to His Excellency. . . . The scene which resulted passes description. He shouted at me in a rage: 'You are the most impudent man in the Bombay Army, Sir!'" The Great Eltchi's anger cowed most men, but not Burton. Months afterwards he learned that his project ran counter to the course of high politics, and hence the fury of Lord Stratford.

Burton was now on the verge of his greatest discovery. At the instance of the Royal Geographical Society he was supplied with funds for an exploration of the Lake Regions of Central Africa, then utterly unknown. In October, 1856, he set out from Bombay, applied for the services of his old comrade Captain Speke, and proceeded to Zanzibar. These explorations were paid for dearly. They cost him twenty-one attacks of fever, temporary paralysis, partial blindness, and a friend. In this place the hardships may be omitted and only the climax described.

The explorers breasted a steep, stony hill and saw below them a streak of light. "I gazed in dismay; the remains of my blind-ness, the veil of trees, a broad ray of sunshine illuminating but one reach of the lake, had shrunk its fair proportions." Was *this* the "great water" of which the natives had told him? "I began to lament my folly in having risked life and lost health for so poor a prize, to curse Arab exaggeration, and to propose an immediate return to explore the Nyanza, or Northern Lake. Advancing a few yards the whole scene suddenly burst upon my view, filling me with admiration, wonder, and delight. Nothing in sooth could be more picturesque than this first view of Tanganyika Lake, as it lay in the lap of the mountains, basking in the gorgeous tropical sunshine. There were precipitous hills, a narrow strip of emerald green, a ribbon of glistening yellow sand . . . and an expanse of soft blue water from thirty to thirty-five miles wide . . . with a background of high, broken wall of steel-coloured mountain flecked and capped by pearly mist."

But now came trouble. Burton believed that in Tanganyika they had discovered the source of the Nile; Speke preferred the view that the great river took its rise in this Northern Lake of which they had heard from the natives. Neither had the slightest real grounds for his belief, but actually Speke's hypothesis was the true one. On the return journey, at Kazeh, the two explorers temporarily parted company; Burton, the linguist of the party, remained behind to organise the return to the coast whilst Speke, equipped by Burton's care, set off to explore the north. In six weeks he returned. He had discovered the Nyanza, the real source of the Nile. During the return to the coast thereafter, Speke grew morose. Then he fell ill of an excruciating disease, and was nursed by Burton through this trouble, from which he emerged peevish, querulous, a changed man. They reached Zanzibar on March 4, 1859. At Aden, Burton collapsed with fever. Speke, abandoning him, took the first ship home, and, despite a promise to Burton, went straight to Burlington House and claimed the whole credit for the discovery of the Nyanza. When Burton, the leader, returned to London, he found the ground completely cut from beneath his feet. Speke had spent his time so well that a new expedition was financed—with Speke in command.*

During Burton's absence in Africa his father had died, leaving him £16,000, a fair fortune; but Burton was always a lavish spender, and much of his capital went in financing his expeditions. In April, 1860, he set off to see for himself in Utah how polygamy suited a European race. He met Brigham Young and inquired if he could be admitted as a Mormon; but Brigham, no doubt thinking of Mecca, shook his head. "No, Captain, I think you've done that sort of thing once before."

On his return to England, Burton married Isabel Arundel. How this appeared to the Burton family may be gauged from the opinion of Miss G. M. Stisted, Burton's niece and admirer. "Looking dispassionately at this match, it is clear that Burton committed as serious an imprudence as when he sent Speke alone to search for the Victoria Nyanza." Burton was forty, Miss Arundel was thirty; neither of them was likely to change. Miss Arundel came of an old English Catholic family and she was bigoted in her faith, putting her trust in miraculous medals, relics, crucifixes, and rosaries. She carried tactlessness to the verge of genius. One gets the impression that she divided her time between adoring her husband and hampering him at every turn owing to her limitations. Not that he himself was very different. He believed in mascots and omens, and carried chestnuts in his pocket to ward off calamity. But there is one thing which must be put to the credit of Burton's wife: she struggled unceasingly to have him recognised for his merits, even

---

* Fuller accounts of the whole matter may be sought in Lady Burton's *Life of Sir Richard Burton*, T. Wright's biography of Burton, and *The True Life of Captain Sir Richard Burton, K.C.M.G.*, by his niece, Miss G. M. Stisted.

if some of her methods may not have recommended themselves to precisians.

At their wedding breakfast, a medical man interrupted a story Burton was telling. "Now, Burton, tell me how you feel when you have killed a man." To which Burton retorted: "Quite jolly, doctor. How do you?" But when the newly-married couple began to count up their resources they can hardly have felt "quite jolly." After he had paid his wife's debts only £4,000 pounds remained to Burton; and this, when invested in a joint annuity, brought in about £200 per annum. In addition to that, there was his half-pay. But here a minor disaster befell. Burton had applied for the Consulship at Fernando Po; he had taken no precautions about his connection with the Indian Army; and as soon as his new appointment was gazetted, he found to his dismay that his name was erased from the Indian Army lists and his half-pay ceased.

Owing to the pestiferous climate of Fernando Po, Burton's wife could not accompany him thither, and he saw her only during leave. Some two years later, however, Burton was appointed by the British Government to be a commissioner and a bearer of a message to Gelele, King of Dahomey; and as a reward for his services in this role he was transferred to the Consulate at Santos, in Brazil, where his wife could join him. Then light seemed to dawn. He was appointed Consul at Damascus, where his familiarity with the Arab seemed to make him at last the right man in the right place. The salary was £1,200 per annum, by no means despicable for those days; and the post was an important one. Thus when he arrived in Damascus on October 1, 1869, his foot seemed at last on the ladder which might lead to higher things.

Two years later he was recalled. A series of unfortunate events had brought him into hostility with almost all the main sections of his jurisdiction. A Copt entered his wife's tent, probably in search of money. Burton's attendants expelled the man and by chance this was seen by a congregation of Greeks filing out from their devotions. They took the part of their co-religionist and a scuffle ensued. The affair was exaggerated in gossip, and thus the Consul and the Greek community were left at daggers drawn. The Jewish moneylenders in the district oppressed the natives, and Burton, by taking the part of the peasants, alienated the Jews. Then arrived a German, a self-appointed missionary, who had not taken the precaution to learn Arabic with accuracy. As a result, he was found beseeching the Moslems to lift up their dog unto the Lord for a broken and contrite dog He would not despise. This arose from the missionary's confusion between *Kalb* (heart) and *Kelb* (dog); but it was hardly an appeal likely to please in a land where the dog is an unclean animal. This German's proselytising activity inflamed the Moslem population, and trouble ensued. The regularly-appointed missionaries lived on good terms with the local Moslems and with Burton, but this German self-appointed fanatic had the ignorance of other races and disregard for their feelings, which mark his countrymen in general. "I should glory in martyrdom," he declared to Burton. "But we should not, nor would the many thousand Greeks and Roman Catholics that inhabit the neighbourhood," the Consul retorted, and he took measures to stifle further trouble. Just in time, for the Turkish authorities were beginning to move in the matter. The German, far from being grateful, hied him to London, where he represented Burton in a most unfavourable light. Another German, a Roman Catholic priest, started more trouble by engineering an hysterical "revival" with tales of visions and rumours of a Second Coming. This in no way concerned the Consul officially; but his wife could not restrain her Roman Catholic fanaticism, and he was drawn into the affair against his will. For the first time in his life he found himself unpopular with Moslems. The Consul-General frowned upon him, and, without a moment's warning, Burton was displaced from his post. At fifty years of age he found himself dismissed like some worthless domestic.

Returning to England, he sank into an abyss of dejection. For ten months the Buttons lived in penury. Once, during a railway journey, they counted up their remaining fluid resources—fifteen sovereigns; and by a mishap, one of the coins rolled on the floor, slipped through a crack, and was gone, reducing them to £14. The scene is described by Burton's wife. "I sat on the floor and cried, and he sat by me with his arm round my waist trying to comfort me."

In 1872, Charles Lever, the novelist and Consul at Trieste, died, and Burton was offered the post. It was worth only £600—half of what he had at Damascus; but beggars cannot be choosers; and thus the world-famous explorer was washed into the backwater where he was to spend the rest of his life. So it came about that the advertisement of a new translation of *The Book of a Thousand Nights and a Night* emanated in due course from Trieste.

What were the special features in which Burton sought to make his version differ from all those which preceded it? They were at least four in number.

In the first place, the original Arabic text utilises in turn the classical language, the colloquial tongue, slang, and the "unprintable," so Burton resolved to vary the vocabulary of his translation to match the original and to pass from standard English into thieves' Latin or Billingsgate where these best reproduced the nature of the Arabic. Secondly, there was the Arabic assonance or "rhyming prose" which Burton endeavoured to indicate by something like this: "And wight fought with wight till the fading of the light and the coming of the night put all foes to flight." Thirdly, there was the verse which constitutes so large a part of *The Nights*, and which confronted Burton with the difficulty of paralleling the mono-rhyme of Arabic:—

"Take warning O proud	*	And in length o' life vain!
I'm Shaddád son of Ad	*	Of the forts castellain;
Lord of pillars and power	*	Lord of tried might and main;
Whom all earth-sons obeyed	*	For my mischief and bain;
Who held East and West	*	In mine awfullest reign."

A gallant effort, evidently, but its success must be judged by those conversant with the original Arabic.

Burton's fourth project was one which he, of all men, was fittest to achieve; and it was the one which brought down the greatest storm of criticism. He proposed to annotate his translation, using it as a vehicle for perpetuating the gigantic store of information which he had acquired about the Arabs, their ideas, superstitions, mode of life, and customs. Now a man may be interested in some things without being vicious himself; and Burton seems to have been one of them. Further, his edition of *The Nights* was printed for private circulation, on the strict understanding that it was not to be allowed to fall into the hands of the public at large. But Victorian modesty was horrified at the mere thought of such matters being discussed at all; and a piercing outcry was raised by many people, some of whom, one suspects, never set eyes on the work itself. The notorious Terminal Essay especially was denounced, and in an article in the *Edinburgh Review*, Burton's version was pleasantly described as "for the sewers." In our days, when the Ministry of Health publishes warnings about V.D. in the advertisement columns of the newspapers, Burton's Notes would hardly excite comment. Even 40 years ago, when I read them first, I wondered what all the fuss had been about. But each time has its own conventions, and undoubtedly his version set yet another black mark to his name among people who knew but little about him.

In order to add to the secrecy enshrouding the actual issue of Burton's *Book of a Thousand Nights and a Night*, the title pages bore the inscription: "Benares. Printed by the Kamashastra Society for private subscribers only." The printing was actually done in England, and "Benares" was a mere jest. The original edition was in twelve volumes and, like Payne's translation, was based on the Macnaghten Arabic text. A comparison between the Payne and Burton versions shows that Burton helped himself pretty liberally from Payne's translation when compiling his own rendering.

But these twelve volumes did not exhaust Burton's store. He informed his subscribers "that my 'Anthropological Notes' are by no means exhausted, and that I can produce a complete work only by means of a somewhat extensive Supplement." He therefore began to issue a further series of volumes—six in all—containing material not found in the Macnaghten text but drawn from the Breslau, Galland, Wortley-Montague, and Chavis and Cazotte versions.

The first two volumes of these *Supplemental Nights* were plain sailing, since the Arabic text of the Breslau version was available; but with his third volume, Burton came on a serious difficulty. Galland's French text contains two of the best-known among all the tales in *The Arabian Nights: Aladdin or the Wonderful Lamp* and *Ali Baba and the Forty Thieves*. But, strange to say, no trace of these stories is to be found in any of the "canonical" versions of *The Nights* in Arabic, such as the Macnaghten text or the Breslau edition. The French text of Galland, and that alone, contained them.

Now Galland, in his dedicatory epistle, declares that his Arabic manuscript was in four volumes; but only three of these have survived, each about 140 pages in length. These do not contain the original of *Aladdin*, and they bring the narrative down no further than the two-hundred-and-thirty second Night. If the missing volume were equal in size to those which have been preserved, it would have brought the work down to the tale of the *Ebony Horse*, which ends with the three-hundred-and-seventy-first Night. But, on this basis, all four volumes of manuscript together would hardly have contained more than two-thirds of the material comprised in Galland's French translation. The extra third, containing *Aladdin*, *Ali Baba*, and eight other tales, must have come from some wholly different source.

What was this source? Eastern manuscripts were known which contained *Aladdin*, but they were merely modern translations of Galland's French text. The suspicion arose that Galland had eked out his original Arabic material with his own invention, just as Cazotte did later; and it was believed by some people that there had never been any Arabic original for these tales. On the other hand there was Galland's character and his honesty as a scholar. It seemed hardly credible that a man of his type would, cuckoo-like, foist off his own offspring into the nest of the Arabian bird. Galland, from all we know of him, was the last man to commit a literary forgery like this.

This was the state of affairs when Burton was engaged in bringing out the volumes of his *Supplemental Nights*, and the problem of *Aladdin* could not be shirked if his collection was to be made a complete one. Some of the subscribers to Payne's translation had complained because it did not contain *Aladdin* and some others of their old favourites.

What was to be done? The obvious thing was to translate Galland's own text. But Galland had gallicised his material to such an extent that any translation would bear the French stamp in ideas even if the phraseology were freed from gallicism; and the result would be inharmonious with the rest of Burton's work. The next possibility was to translate Galland into Arabic and then retranslate this version into English. This process, however, when tried, was found wanting.* Burton then fell back upon a Hindustani version of the Galland tales in which they were sufficiently Orientalised and divested of their inordinate gallicism. This Hindustani version was translated into English, as that seemed the best that could be done.

But at this juncture, by a coincidence worthy of *The Arabian Nights* themselves, an event occurred which simultaneously cleared Galland's credit and gave Burton exactly what he needed. During a visit to the Bibliothèque Nationale in Paris, he met M. Hermann Zotenberg, Keeper of the Oriental MSS. in the French national

---

* This calls to mind that peculiarly ludicrous attempt to write a great Irish drama which should be free from the slightest trace of English influence. It is chronicled in the first volume of George Moore's *Hail and Farewell*. George Moore was to compose the drama in French; Lady Gregory was then to translate Moore's French into English; Taid O'Donoghue (who spoke Gaelic) was next to turn Lady Gregory's English into Irish; after which Lady Gregory was to render Taid O'Donoghue's Irish back into English; and, finally, W. B. Yeats was to polish up the style. This laborious process was expected to yield something free from any English taint and racy of the Ould Sod. It seems a pity that nothing came of it, owing to disagreements among the numerous collaborators.

collection; and he learned that not long before this, M. Zotenberg had purchased a manuscript copy of *The Nights* containing the Arabic originals of the tales of *Aladdin* and of *Zayn al-Asnam* which Galland had issued in his translation. Examination showed that this MS. had been written in A.D. 1703; and as Galland did not begin the publication of his translation until 1704-5, it was plain that the Arabic MS. could not be a translation of Galland's version. This discovery established that Arabic MSS. of these two tales existed before Galland's work saw the light and thus fell the charge against Galland that he had invented the stories himself and passed them off on an unsuspecting public as of Arabic origin. Thus, though no originals of the eight other tales—*Ali Baba and the Forty Thieves* amongst them—have yet been unearthed, it is safe to reckon that Galland had Arabic manuscripts of them in his possession, though these are now lost.

Burton brought out the first volume of his version of *The Book of a Thousand Nights and a Night* in 1885, when he was sixty-four years of age. Up to that point, his life had been a series of disappointments. Brilliant feats of exploration had met with only inadequate recognition. In the case of the Tanganyika expedition he had seen the credit filched by a subordinate. Twice he had taken his life in his hands in the Government service: in the expedition to Harar and in the mission to Gelele. In return, not even the lowest decoration had come his way. The Damascus Consulship could raise only bitter memories. The Tangier post on which he had set his heart was denied him. He had known penury. His books were never great successes. And now, within five years of his death, he was stranded in an unimportant post in Trieste which gave no scope for the talents he felt that he possessed. Small wonder that he was angry and embittered.

But at last Fate relented. His version of *The Arabian Nights* brought him £10,000; small wonder if he spent that lavishly. And, during the appearance of his volumes, the Government at last decided to honour him, for in 1886, at the age of sixty-five, he was created a K.C.M.G.

His life had been tempestuous, and even as he lay awaiting sepulture, fresh storms were gathering. He died suddenly, quite insensible at the last, under the great map of Africa with its Arabic motto: "All things pass," which hung at the head of his bed. Immediately his wife summoned "a country Slav priest" and begged him to administer Extreme Unction and so assure Burton's salvation. The priest demurred. Both he and the doctor believed that life was extinct, so that it was too late for any death-bed conversion. But Lady Burton would take no denial.* "He is alive," she cried, "but I beseech you, lose not a moment, for the soul is passing away." "If he is a Protestant," said the priest, "he cannot receive the Holy Sacrament in this way." Thereupon Lady Burton declared that her husband "had abjured the heresy and belonged to the Catholic Church," and on the strength of this assurance, the priest consented to perform the rite.

These are deep waters, into which it would be unsafe for a layman to enter; but the results in this world were not left in doubt. Lady Burton was consoled. Burton's own family were deeply offended. "In spite of numerous and pressing invitations (to the funeral) only one member of the (Burton's) family, a distant cousin, accepted: sister, niece, his favourite relatives, and many of his best and most sympathetic friends, refused to countenance a Lie."†

And what would Burton himself have said? One can but conjecture. Possibly he would have laughed that peculiar laugh of his, "like the rattle of a pebble across a frozen pond." Nothing less was likely from a man who, in his own words, "found four great Protestant Sommités: St. Paul, who protested against St. Peter's

---

* Some months later she admitted that Burton was dead before the administration of Extreme Unction took place. See Wright, *Life of Sir Richard Burton*, Vol. II, p. 244 (1906), for an authoritative account of the whole episode.
† These words indicate the feelings of Burton's niece, Miss G. M. Stisted, and are quoted from her *True Life of Captain Sir Richard Burton, K.C.M.G.*, see p. 415.

Hebraism; Mahommed, who protested against the perversions of Christianity; Luther, who protested against the rule of the Pope; and Sir Richard Burton, who protested against the whole business."

Having settled the matter of her husband's soul, Lady Burton turned to deal with his literary remains. At the end of his life, Burton had set himself to translate and annotate an Arabic work: *The Scented Garden* of the Shaykh Nafzawi, which he proposed to issue, for private circulation only, among 1,500 selected subscribers who had already enrolled themselves. The manuscript of this translation was among the papers which Lady Burton set herself to examine. She was horrified, and, considering her mental equipment, no one can wonder at that. It was hardly her kind of book. She might have consulted someone expert in the subject; but that was not her way. Instead, she took the advice of the local peasant priest; and when this was confirmed, apparently, by three visions of her husband's spirit, she put the whole manuscript into the fire without more ado.

This raised a fearful outcry from the 1,500 would-be subscribers, who were thus disappointed. Whether the loss of the actual translation was a serious matter is open to doubt. Anyone who has read the Burton-Arbuthnot edition of the *Ananga-Ranga* may be excused for thinking that *The Scented Garden* could hardly give him much further enlightenment about Oriental amatory exercises. In any case, both French and English translations of the text had already appeared. But Burton's anthropological notes were certainly irreplaceable and were probably valuable. Lady Burton, however, did not deem them so, and they went into the flames along with the text.

Whatever one may think about this destruction of a work on which her husband was engaged during his last days, there can be no two opinions about Burton's diaries. They, undoubtedly must have contained much that was of value, for he seems to have kept them with care; and the journal of such a man should have been preserved. But Lady Burton had no qualms, and she destroyed the diaries also.

She devoted herself to the erection of a marble tomb at Mortlake, an imitation of an Arab tent, and in it Burton's embalmed body was preserved: a curious haven for the remains of a man who was probably the greatest combination of linguist, anthropologist, and traveler of his day. On the anniversary of Burton's death she each year "dressed the mausoleum." In 1893, she wrote to the wife of Burton's cousin a letter describing the ceremony, which has been quoted by Thomas Wright in his *Life* of Burton. "We had mass and communion," she explains, "and crowds of friends came down to see the mausoleum and two photographers." One hopes that the photographers were flattered by this attention.

VII
WHAT IS YOUR EVIDENCE WORTH?

"I SAW JONES IN THE FIELD, about 200 yards away. Smith was 20 yards nearer me. He made at Jones to give him a clout on the jaw. Jones ups with his rifle; pulls the trigger; fires two shots, one atop of t'other; and Smith drops down dead. When I ran up, Smith said something I didn't catch. Then Jones said something, but I don't remember what it was. That would be about four, or maybe half-past four o'clock; I can't say nearer, for I'd no watch on me."

Now, suppose you were faced with that evidence, how much of it would you accept as accurately based on observation?

In the first place, the "clout on the jaw" is pure surmise. All that the witness can have seen was Smith making threatening gestures as he approached Jones. Further, at a distance of 200 yards, it would be quite impossible for the witness to see Jones pulling the trigger. That is not observation, but an inference from the fact that the rifle was discharged. Again, Smith did *not* drop down dead. That is a verbal frill, an attempt to be graphic; for afterwards the witness reports Smith as speaking. Moreover, only one shot was fired. Anyone who has stood in the butts on a range will recall the double report "klick-klock!"—made by a rifle when fired toward an observer. This is due to the separate arrivals of the percussion-wave and the sound-wave from the weapon, one of which travels faster than the other. Furthermore, there is a lacuna in the evidence, due to the witness failing to catch what Smith said; and there is another lacuna left by his forgetting what Jones remarked. And, finally, the time of the occurrence is estimated, since the witness carried no watch. If he is a townsman, his estimate will probably be astray; whereas if he is country-bred he will be accustomed to judging time from the position of the sun, the length of shadows, etc., and his guess will probably be a fairly accurate one.

From this imaginary example, it is clear that the value of evidence volunteered by an honest witness can be properly assessed only if we have some understanding of various factors which enter into the problem.

All evidence from the outside world comes to us through our senses,* and the perfect witness would be one who perceived every sensation, recorded them exactly as he experienced them, and was able to describe them precisely, when his testimony was required.

Within its mechanical limits, we have such a witness in the instrument used to produce the coloured talking-film of the cinema. Imagine such a machine set up in a village street and fitted with a chronograph to record the exact instant at which each exposure was made by the shutter of the camera; what would be the result. If Mr. Robinson came out of the village public house and crossed the street to a tobacconist's shop, singing as he went, we should have an exact record of his behaviour in transit, from which we might infer some things about his state of mind and his degree of sobriety, simply by completing the usual process and throwing the resultant positive pictures on the screen by means of a projector. The evidence thus obtained would be wholly unbiased and would be complete in every detail, so far as light and sound were concerned. In other words, the instrument would be a perfect recorder of exact evidence about Robinson during the time between his

---

* It is usual to speak of our "five senses," but actually we have six. We can distinguish between hot and cold water by dipping our hands into it; and in each case we are dealing with water, so we cannot be depending on the ordinary sense of touch in this case, but must have a special heat sense in addition to the other five.

emergence from the public house and his vanishing into the tobacconist's shop.

But now suppose that the camera is in charge of a less responsible person. While he is busy recording the movements of Robinson, the village belle comes down the street. The photographer, if he be susceptible, will swing his machine round to take a few pictures of her, and meanwhile Robinson's doings go unrecorded. Or some rain may drive on to the lens, and blurring of subsequent exposures may result until the moisture is wiped away. Again, the film in the camera may be faulty, and some of the exposures may fail to produce any record at all upon it. Further losses of this kind may occur during development and printing. Then imagine that this faulty film is handed over to someone for cutting. Naturally, the cutting will vary according to the taste of the cutter. A policeman would concentrate upon those parts of the strip which showed Robinson giving a song-and-dance turn in the middle of the road, since that would be evidence against him on the charge of being drunk; a small boy, if put to cut the film, would probably concentrate on the ludicrous parts—Robinson knocking off his hat and trying to recover it, or tripping over the kerbstone and falling flat, etc.—whereas a motorist would be inclined to cut the film in such a way as to show the difficulties which Robinson's behaviour puts in the way of traffic. By the time the cutting had been done, the film would have ceased to be an impartial record, but would have turned into something rather like the testimony of an honest witness struggling with difficulties raised by faulty observation, failure of memory, and his own innate prejudices.

Before a witness can give an account of something which he has experienced at an earlier period, various phenomena must occur. For the sake of clarity, it will be well to mention them in their proper order at this point and deal with them in more detail later on.

In the first place, the witness receives a series of *sensations* from the outside world. The vast majority of these are completely ignored by the witness; but some of them succeed in imprinting themselves on his *perception*, especially those which are likely to be of value to him for his ultimate advantage. If his interest is aroused beyond a certain degree, his power of *observation* is stimulated and he pays more attention. Sometimes he is not content with observation *per se* but proceeds to place an *interpretation* on the results. The next process consists in storing these things in his memory by recording the experiences he has undergone. Once in his memory, however, the data inevitably suffer a partial *effacement*, in which certain of the original material vanishes from his mind. If, now, he is asked for his testimony, he is apt to make a second process of selection or *excerption*, rejecting whatever may seem to him irrelevant, instead of pouring out all that he can actually recall. Here he comes to the difficulty of putting this excerpted material into the form of a clear statement and so making a proper *presentation* of his evidence. Finally, if he tells his tale several times, his testimony is apt to undergo what is termed *crystallisation*, taking on a more and more definite form which was lacking in the initial presentation.

Taking a concrete example. A man, musing over his love affairs, goes along the street in a brown study, paying no attention to passers-by, though his perception suffices to prevent him from jostling them on his way. Coming to a newsboy, he recalls that he has a bet on the three o'clock race and he hands over a coin, gets a paper, and receives change. Later, he finds himself with less money than he supposed he had. This observation leads him to the interpretation: "That boy gave me the wrong change." Next day, in telling the story of these events he says: "That boy cheated me, though perhaps he made a mistake." A week later, after several tellings, this crystallises into: "That boy's a cheat."

Generally speaking, the mere act of perception is unlikely to yield evidence of much value. In walking along a crowded street you obviously perceive a man who comes towards you, since you avoid colliding with him; but you would probably be unable to describe, later on, the appearance of a casual passer-by. But if one of these strangers stops you to ask his way, your attention is directed to him in particular and your power of observation is called into play, so that, later, you might be able to identify him if necessary.

Observation itself, however, is subject to marked defects. During the 1914-18 war, one of the "catches" in vogue depended on this. A diddler, after a drink or two with his dupe, placed his tumbler, mouth downwards, on the table. He then borrowed a florin from his victim and placed this florin on the uppermost surface of the tumbler. "Now," said he, "would you say that the florin was on the top or the bottom of the tumbler?" The dupe would consider the matter and give his verdict: "Top" or "Bottom." Whichever his choice, the diddler looked disappointed, took up the florin, saying: "Oh, I see you know the catch." Whereupon he handed back the florin to his victim, who pocketed it, feeling that he had shown himself astute in penetrating the trick. If he had observed more carefully, he would have seen that what the diddler actually handed to him was a penny instead of the original florin; but as he received a coin of the same *size* as the florin, the dupe never troubled to look at it but thrust it, without closer examination, into his pocket. Net profit to the diddler, 1s. 11d.

The French criminology expert, Dr. Locard,* quotes two instances in which perfectly honest witnesses made statements which were demonstrably erroneous. In the one case, a witness declared that while standing at a certain street corner he had seen the accused man emerging from a side alley. Examination of the ground showed that the exit from the alley was invisible from the point where the witness had been standing. Again, several witnesses accused a priest of committing acts of indecency while in the confessional; but investigation proved that from where they were standing it was impossible to see into the confessional.

Illusions of the senses may enter into the problem of observation. Thus, when a witness looks along a straight stretch of railway, his eyes show him the pairs of rails converging, whereas his reason assures him that they are parallel. When he is on the deck of a ship which is leaving the quay, the motion may be so insensible to him that it appears as though the quay were the moving

---

* E. Locard, *L'Enquéte criminelle*, p. 32.

object receding from the ship. The same kind of thing is seen when a passenger in a train feels the station sliding away behind him as the train moves out of it. In these cases, reason overcomes the illusion quickly enough; but unless reason has a very firm basis, the illusion might remain predominant in the mind of the witness. For example, everyone knows the "cog-wheel" diagram which is employed to check the speed indicator of a gramophone table. Seen by daylight, the spin of the cog-wheel is easily detected. Now, to the ordinary person, the light from a 50-cycle A.C. electric circuit appears to be continuous illumination; but this is mere illusion, since actually it consists of a series of flashes with dark intervals between. If the speed at which the "cog-wheel" diagram revolves with the gramophone turntable is arranged so that during the "dark interval" in the electric lighting the cogs in the diagram have time to move forward one place, then despite the revolutions a witness would say that the "cog-wheel" remains steady and motionless. Here the average person does not understand what is happening, and if he were shown the phenomenon without further explanation he would be quite prepared to swear that the turn-table of the gramophone was not in motion, since his illusion would not be corrected by his reason. Again, an illusion may receive support by blending it with reality in such a way that our recognition of the real things tends to make us more receptive of the illusion. Thus, if a painter of a panorama wishes to enhance its resemblance to a real landscape, he will place real stones and grass in the near foreground and make his painting merge into these actual objects if possible. The eye recognises the stones, etc., as real, and thus the mind is led to accept the picture further back as a real moor, for example.

Hearing is another sense which is liable to considerable error. From the point of view of physics, the intensity of a sound heard by a witness depends upon two factors: first, the distance of the witness from the generator, and second, the intensity of the sound produced by the generator. Thus, a loud sound in the distance may easily be confused with a nearer but softer sound, provided the two are of the same type, such as explosions.

In legal affairs, a good deal depends at times upon the ability of a witness to recognise the voice of a person. What is needed here is the appreciation of three elements: pitch, intensity, and timbre; and, further, a mental comparison with the same three factors in the known voice of the suspected person. Even if the witness is perfectly familiar with the ordinary speaking voice of the suspect, he may fail entirely to recognise the same voice speaking in a whisper, since in the latter case all the three factors are modified. Anyone can test this for himself by asking three friends to converse in whispers in a room while he himself stands listening behind an open door.

Touch is another sense which is liable to give rise to mistakes. If a grain of dust gets into one's eye, it appears quite large, and when it is extracted and examined, the usual feeling is: "Was that tiny thing the cause of all the trouble?" Again, if a, seed gets wedged between one's teeth, the tongue perceives it as large, and again when it is extracted and examined, some surprise is caused by the minuteness of its real dimensions. Touch is subject to illusions also. Everyone will recall the old experiment of crossing one's middle finger over one's forefinger and inserting a pencil or a pea into the V formed by the crossed fingers. Two objects are perceived instead of the single one used. Moreover, we depend so little on touch for ascertaining the shape of things that we go far astray if we rely on tactile sensations alone. The reader can convince himself of this by a simple experiment. Let someone blindfold you efficiently and then hand you some object such as a rather complicated china ornament, a wooden golf-tee, a spring letter-balance, or one of these machines for embossing an address on notepaper. Attempt to make a guess as to the material, bulk, and form of the object, and then try to put your results into words. After you have tried this, you will have enough experience to estimate the value of the testimony of a witness giving evidence depending upon his sensation of touch in the dark.

So far, we have been considering the case of a witness with normal senses; but it may well happen that the witness's senses suffer from defects. Suppose that in the case of an assault, the complainant's only recollection of his assailant is that he wore a red tie. A confirmed Communist is suspected and it is known that his habitual neckwear is a red tie, so the case looks as if it were strengthened by the witness's evidence. But now assume that the witness himself is examined thoroughly. It may be found that he suffers from colour blindness or, as our medical friends would call it: achromatopsia. As a result of this defect of vision, the witness "sees" green as red, so the assailant may really have been wearing a green tie and not a red one, at the time of the fracas. This suggests that no witness's testimony as to colours should be received until after he has been medically examined and the possibility of colour blindness has been eliminated.

In the case of hearing, Locard* describes a case of some interest. A witness averred that a certain event occurred in a mill at noon precisely. How did he know the exact time? He declared that while the event took place, he heard the chimes from the nearest steeple strike 12. An independent test was made, and it was found that the noise of the mill entirely drowned the sound of the chimes. The witness then recalled that in reality he had heard the hour strike *after* he had emerged from the mill. Yet this testimony might have been perfectly accurate if he had been a sufferer from a certain form of deafness. In this particular variety of surdity, vibration improves the impaired hearing, so that the patient, when in a bus or a tube, hears better than when in a still place. Such a patient, placed amid the grindings and shakings of the mill, might quite possibly pick up something which would be inaudible to the normal ear, deafened as it would be by the surrounding racket.

Another factor influences our sensations: emotion; and it seems possible that too little attention is paid to it in some cases, especially in cases of criminal violence when the witness is probably highly-strung. Two examples will suffice. The professor of Criminology in Prague University, Gross,† recounts how he witnessed

---

* Locard, *op cit.*, p. 31.
† Gross, *Criminal Investigation*, p. 52, Third edition, 1934.

an execution during which the executioner wore gloves. After the execution, Gross asked four officials who were present: "What was the colour of the executioner's gloves?" Three replied, respectively: black, grey, and white; the fourth stoutly maintained that the executioner wore no gloves at all. All four were close to the scaffold; all four replied without hesitation; and all four were absolutely confident that they had made no mistake. Obviously, the nerves of them all had been strung up by their situation, and their normal powers of observation were impaired.

Gross also cites the case of the execution of Mary, Queen of Scots. When, between 1830 and 1840, her coffin was opened, it was discovered that she had received not one but two strokes of the axe, the first of which merely slashed the nape of her neck, whilst the second severed her head from the trunk. Now there are numerous contemporary accounts giving an abundance of precise details, yet not one of these mentions the first stroke which failed in its effect.*

A third case is mentioned by Gross: the assassination of President Carnot by Caserio Santo in 1894. The murderer jumped on to the foot-rest of the carriage, pushed the President's arm aside, and drove a dagger into his abdomen. In the carriage with Carnot were three gentlemen; two lackeys were standing behind; mounted officers were riding on either side—yet not one of these saw the fatal stab. In fact, the murderer might well have escaped undetected if he had not, with the theatricality of the Latin races, shouted out: "Vive l'anarchie!"

Turn next to another aspect of the matter. The testimony of a witness often increases in value if he can give a good estimate of time or of distance. As has been mentioned already, a country-bred man will probably make a closer guess at the time of day than will a townsman, if both are deprived of the aid of a timepiece. As for

---

* Brantome appears to be the only writer to mention more than a single stroke; but he errs by excess and asserts that no less than three strokes were made by the headsman.

estimating the lapse of time between two events, a doctor with practice in counting pulses is more likely to be correct than a layman, when it is a matter of seconds; and a man who has learned the trick of saying: "One *Thousand,* Two *Thousand* . . ." when counting seconds is much more likely to be near the mark than anyone who counts simply: "One, two, three . . ." for the latter is almost certain to count too fast, and so over-estimate the actual interval.

The gauging of distances is another matter in which witnesses vary considerably in aptitude. For example, an amateur runner will have a fair idea of 100 yards, 220 yards, a quarter-mile, etc. A golfer would say to himself: "It would take me a drive and a mashie to cover that distance," or: "I would have taken a brassie to it"; and as he knows the average length of his various strokes, he could make a fair estimate. A fresh difficulty appears when a witness has to judge distance across intervening water, say from one side of a small lake to the other, since in such a case there are no intervening objects whose sizes might furnish a gauge. Here a seaman, accustomed by his profession to the appearance of water, would probably be more accurate than a landsman.

So much for that part of the subject. But when a witness has observed a phenomenon he must then store his recollection of it in his memory or his testimony will be void. Here selection enters the problem; for he will remember mainly the things which interest him and will forget the rest. Now many facts which are of the most vital importance from the legal point of view may have not the slightest personal interest for the witness. In fact, they may seem to him of so little importance that he does not trouble to observe them closely, being more engrossed with matters which have absolutely no value whatever in the eyes of the law.

Each witness brings his own idiosyncrasy. For instance, if before a certain landscape we place in turn a farmer, an artist, an architect, and a botanist, each of them will regard the scene from the stand-point of his ruling interest. The farmer will see it from the agricultural point of view; the artist will consider it pictorially; and so for the others also. If a number of witnesses pass along a street and inspect the shop windows, they are unlikely to pause at the same places. A man with literary tastes will stop at a bookshop; another, who makes a hobby of his workshop at home, will halt before a window full of tools; whilst a woman might ignore both of these and be attracted instead by a display of hats.

Moreover, even when witnesses observe the same thing, each comes to it with his own prepossessions. In a street, a man is likely to give a pretty girl more attention than he would pay to a nondescript. He will, in all probability, take note of the girl's general appearance, her looks, her figure, the neatness of her legs and ankles, her carriage, the way she walks, and so forth. A woman witness, following in his track, would probably notice similar points; but if she were questioned afterwards, she would very likely be able to remember much fuller details about the girl's dress than the male witness could recall.

It has already been pointed out that sometimes a witness is not content with merely storing up the impressions which he has received. Before recording them in his memory he may try to put an interpretation upon them, and quite possibly by so doing he deforms the actual evidence of his sensations.

For instance, our experience tells us that the further away from us an object is, the less clearly do we see it. Now the interposition of a screen of mist blurs the details of an object without altering its distance; but owing to the blurring we are apt, unconsciously, to think of it as more remote than it really is. As the length and breadth of what we see remain as before, we tend to exaggerate its size and to imagine that it is larger than it actually is in reality, but that it is further away from us than it really is. This accounts for the well-known fact that during a naval action in a haze, an enemy destroyer close at hand is sometimes mistaken for a cruiser steaming at a greater distance.

Occasionally a witness may put one interpretation on the sensation he has received; then, dissatisfied with this, he may discard his initial interpretation and replace it by a second. The classic case of this is recounted by Locard* from his own experience. One

---

* Locard, *loc cit.*, p. 38.

night he was sitting with a man at a spot between a frog pond and a road. A distant sound came to their ears, whereupon Locard's companion remarked: "That dog isn't a frog; it's the wheel of a cart." It is easy to see from this what happened. The man heard a vague sound. Instead of being content to record this simple fact in his memory, he began to interpret his sensation. "That's a dog barking." Then, on further consideration, he grew dissatisfied with this view, and found another explanation: "No, it isn't a dog; it's a frog croaking beside the pond." Again he considered, and now he reached the conclusion that the sound was caused by a vehicle's wheel on the road near by. The whole process of reasoning was so swift that it took place while he was uttering the sentence: "That dog isn't a frog, it's the wheel of a cart." And in recording the incident in his memory, he would probably ignore his difficulties of the moment and store up an impression that he had heard the sound of a cart wheel—which may have been accurate or inaccurate in reality.

Another peculiarity of the mind is its tendency to group sensations together and then record the group as a whole, discarding the separate sensations which led up to the result. Thus, when we look at the picture on a cinema screen, our reason tells us that we are witnessing a succession of isolated photographs in rapid succession. None the less, we "see" the film as a representation of things in continuous movement. Our habit of mind leads us to fuse the individual pictures into something which is represented by none of them separately; and if we are afterwards asked to describe what we have seen, we launch into an account of the plot of the story, ignoring entirely the individual "shots" out of which it is built.

A rather simpler example of this "grouping" tendency is to be found in cards and dominoes. When an opponent puts down, say, a nine of clubs, it is not necessary for us to count the pips on the card. In fact, we never do so for we recognise the value of the card merely by a glance at the general arrangement on its face, and not by the number of the pips displayed. But if the same number of pips were arranged at random, we should be reduced to counting them one by one, so far as the higher cards go. The same holds good for dominoes. We know the double-six at a glance; but if twelve spots were placed on the same area in random positions, we should have to add them up, unit by unit, in order to recognise the value; and that would take a longer time.

If a man be a quick reader, he carries this "grouping" tendency into that field also. A child learns to read by observing each letter in turn and then synthesising the results mentally into the word which the symbols express. But as his facility increases, the learner tends more and more to look at the appearance of the printed word as a whole, instead of taking it letter by letter.

An illustration of this is to be found in proof-reading. The misplacement or omission of a letter in a short word is detected immediately, because the change of a single letter in a short word makes a very marked alteration in its general appearance. Thus, if *but* is misprinted as *bat*, we perceive the error immediately. The mistake of *concieve* for *conceive* is also readily remarked, owing probably to the change in position of the dot of the "i" which lies above the run of the type and is thus very noticeable. But in the case of longer words, the omission or change of one letter makes far less alteration in the general appearance of the print and is thus not so easily seen at a glance, as in *tintinabulation*, *subserviance*, and *juresdiction*.

Another of the "catches" current in the last war demonstrated this curious point very neatly. If the reader will examine the carton of a packet of "Gold Flake" cigarettes, he will find on it the legend: "Every genuine 'Gold Flake' Cigarette bears the name W. D. & H. O. WILLS"—the name of the firm being in script whilst the rest of the lettering is in ordinary type. Let him hand the carton to an ingenuous friend and ask the victim: "How many aitches can you count in that part of the design?" The result will probably surprise the reader; for it is astonishing to notice how many people answer: "There's only one H there." There are actually two aitches: one in the name of the firm and the other in the word "the"; but the average person misses the latter H completely. This is due to the fact that most educated people, in reading, treat the word "the" as a group and pay no attention to its individual letters, so that the H here escapes notice. But if you try the experiment with a child in the early stages of its reading, it will not make the same mistake, because it is still at the point where it examines every letter in turn instead of reading the word by general appearance.

In this connection, another experiment may be tried. Choose a book printed in fairly large type. Take a post-card and lay it on a line of the printing, so that the upper half of each letter in the line is covered. Now try to make out the sense from the exposed lower half of the words. You will find it troublesome. Try again, but this time cover the lower half of the letters with the card; and you will find it simple enough to grasp the sense. The difference in ease between the two cases can hardly be due entirely to the projections above and below the line of various letters; for although eight of them (b, d, f, h, i, k, l and t) rise above the ordinary level, no less than five (g, j, p, q and y) have characteristic "tails" extending below: and the ratio 8:5 is hardly large enough to account for the result in reading. One is apt to infer that in reading we pay special attention to the upper parts of a letter and tend to ignore to some extent its lower configuration,

Locard* gives from his own experience a quaint error in the interpretation of observations. In his ambulance, attached to the Fourth French Army in the 1914-18 war, there stood an aneroid barometer which, in general appearance, resembled a clock. At average barometric pressure the needle stood near the vertical. Visitors came in, glanced at the dial, read it as they would read a clock, and were almost invariably led to assume that the time was noon. Locard states that if these visitors—and they were many—had been called as witnesses, they would have been unanimous in timing their visit from this erroneous inference that the hand of the barometer was the hand of a clock pointing to twelve.

When we come to the question of sound this "grouping" tendency assumes especial importance. If the reader will refer to the evidence of Melin and Bataillet in the Chantelle case, which is

---

* Locard, *op. cit.* pp. 41-42.

described elsewhere in this volume,* he will find an excellent example. Melin, hearing certain sounds, put his own interpretation on them and then, in accordance with his preconception, he grouped the sounds into a series representing the progress of a struggle between a man and some assailants. Actually, as Bataillet's evidence established, the noises came from men pushing a truck, so that the real series was different from Melin's imaginary one.

The hearing of conversation is an experience so common to us that usually we take the underlying process for granted and do not realise how extremely complex it actually is. Suppose that a friend says to us: "I thought I saw a rabbit in the Castle garden, but it turned out to be a cat." In the first place, our friend emits a certain series of sound waves to express his meaning. These waves reach our ears in the form of vibrations, which are picked up by our aural apparatus. Then our intellect gets to work and transmutes the vibrations into the sensation which we call sound; and in doing so it cuts the message up into its component words. Again the mind operates and replaces the words by pictures—our friend, a rabbit, the Castle garden. Then comes the second clause, and we have to replace our mental picture of a rabbit by another mental picture of a cat. And, finally, we get an idea of what has happened, whilst in all probability the exact *words* emitted by our friend are not remembered, since they have served their purpose in conveying the *idea* which is, to us, the important factor.

That represents roughly what happens when the auditor hears clearly every word spoken. But if the speaker be far off, some of the words may be lost. He may, for example, be trying to shout across a river on a gusty day, and only fragments of his sentences reach his auditor. In that case, the auditor will try to fill in the gaps from his imagination as best he can; and if he were later questioned about the message, he would probably furnish a version based not on what he actually heard but upon the fragments pieced out with fresh matter due to his own conjectures; and thus the

---

* See Chapter 1.

original speaker may be credited with things which he never actually said.

At times, the mishearing of even a single word may alter the whole aspect of affairs. This is well illustrated by one of the incidents in the so-called Green Bicycle Case of 1920, which arose out of the discovery of the dead body of a young girl, Bella Wright, who was found on a road near Leicester with a bullet-wound in her head and her bicycle beside her.

The accused man's story was simple. While out cycling on 5th June, 1919, he chanced on Miss Wright, whom he had never seen before. She was in trouble with her machine, so he lent her a spanner; and when the repair was completed, the couple cycled on to Gaulby. Here they parted. The girl went into her uncle's house, saying to the accused that she would be only ten minutes; whilst the accused man turned and rode back towards Leicester. Being delayed by a puncture, he rode back towards Gaulby, expecting to meet Miss Wright. But she had stayed longer than she intended, and he encountered her again, just as she was coming out of her uncle's house. That was about 9:15 p.m. Joining company, they returned towards Leicester until, not long afterwards, their ways diverged and the girl left him. Shortly before ten o'clock, the girl's body was found in a lane near Little Stretton, about 2½ miles from her uncle's house.

Now, on the face of things, it is most improbable that a man would murder a girl on an acquaintance so slight that he does not even know her name. If—as he stated—he met her for the first time within two hours of her death, it would require the strongest evidence to implicate him in the matter; for no probable motive could be adduced. On the other hand, if he were an acquaintance of some standing, various possible motives might be suggested. Thus, one of the main points in the case was the proof, or disproof, of a previous acquaintanceship between the man and the girl.

The girl's uncle testified that he saw his niece meet the accused man as she was leaving the house in Gaulby. According to him, the accused greeted her with the words: "Bella! I thought you'd gone the other way." The prosecution laid great stress on this, because the use of the girl's name might be construed into an indication of acquaintanceship. The uncle refused to diverge from his statement, although he frankly admitted that his niece had declared at the time: "The man is a perfect stranger to me."

For the defence, Sir Marshall Hall fastened on the first word. Was it not possible, he asked, that what the accused man said was really: "Hello, I thought you'd gone the other way." The uncle denied this. The accused man, when called as a witness, declared that he said: "Hello! You've been a long time. I thought you'd gone the other way."

Eventually the accused was acquitted; but one is left to surmise what might have happened if the jury had credited the uncle's version and disbelieved in the alternative "Hello!" which, on the face of things, concords neatly with the girl's own statement to her uncle.

Enough examples have now been given of the manner in which interpretation may deform an impression left upon a witness by the phenomena which have affected him. It is time to take up the next stage, in which this impression is stored up in the memory.

Locard* likens this stored-up image to a delicate pastel. So long as the pastel is kept in some dark place of safety, it will retain its pristine freshness; but if, on the contrary, it is often brought out into sunlight, handled continually, and exposed to the consequent risks, it will deteriorate bit by bit until it has lost all its original brilliance and colour. In the same way a memory, stored up and seldom recalled, will retain much of its original sharpness; whereas if it be consulted again and again, it loses at each reference something of its previous value.

One method of impressing a thing on the memory is by reinforcing one sensation by another. Thus, if one has to look up a telephone number and has to remember it for a short time before one can dial it, the visual memory may be reinforced by repeating the number aloud so that the ear also gets its chance of making a record as well as the eye.

---

* Locard, *op. cit.*, pp. 46 *et seq.*

It is not always sufficient to record actual words, for, in certain cases, the intonation of speech may be almost as important as the language. Sir Marshall Hall had an uncanny knack of using this in some of his cases, for it is a dangerous tool. In his first important case, a prostitute, Marie Hermann, brought a man home with her; a struggle was heard in her room; and eventually the client was found dead, having been beaten about the head with a poker. A witness testified to hearing the woman Hermann say: "Speak! Speak!! Speak!!!" the word being louder each time and the final repetition being almost a shriek. Here, obviously, the crucial point is the intonation. Were these words spoken in argument and rage? Or were they expressions of pity? Or, again, was the woman persuading herself against hope that her paramour was still alive and trying to get him to make some sign of life? By such suggestions, Marshall Hall succeeded in influencing the jury into making a strong recommendation to mercy, and the woman got off with six years' penal servitude.

In another case he employed similar ingenuity and was even more successful. In 1922, Parliament passed the Infanticide Act under which a mother who kills her new-born baby before she has fully recovered from the effects of childbirth is not guilty of murder; but this case falls into an earlier period when poor women whose trials had temporarily disordered their minds were none the less charged with murder if they committed infanticide. A young girl, Annie Dyer, was seduced by a married man and gave birth to a child. Some ten days later, she left the lodgings where she had borne the baby and arrived at her married sister's house without making any reference to the child. She was traced and a police inspector took a statement from her. In it she said, according to the official: "I will tell you the truth. I killed it—I did not know what to do with it—I put it in a box; you will find it there."

Now the whole case hinged on one main question: had she or had she not acted of malice aforethought? Or, in other words, had she pursued a carefully thought out plan to get rid of an unwanted child? Evidence was brought by the prosecution to prove that ten days after the birth, while sitting with the monthly nurse and the baby, the young mother asked the nurse: "How can anyone get rid of a baby?" Shortly after this, she took the baby to bed, and it was never seen again alive.

The phrases quoted above seem to put the matter beyond doubt. Take them literally and they suggest nothing less than a deliberate killing of the child. But Marshall Hall was able to put a wholly different complexion on the case. Suppose a change of intonation and we get the question: "How *can* anyone get rid of a baby?" Put in this way, the inquiry is a rhetorical expression indicating such fondness for her child that she considers that no mother could even countenance for a moment the possibility of doing harm to a baby. The fact that at the moment when she spoke, Annie Dyer was kissing and feeding the child, lent strong support to Marshall Hall's argument.

The second set of phrases Hall explained in a different manner. Punctuation is no part of spoken language. In this case it had been inserted by the police inspector. But suppose that his punctuation were mistaken? Then, instead of linking the phrases as the inspector did, one gets this result: "I killed it. I did not know what to do with it; I put it in a box," and this would imply only that she over-lay the child (accidentally, perhaps) and then, staggered by what had happened, she "did not know what to do with it" (the body). This is very different from the suggestion produced by: I killed it; I did not know what to do with it," which carries with it the implication that she was in such a fix that she had to rid herself of the living child, even at the cost of killing it.

The jury accepted Marshall Hall's readings, and without even retiring from the box, they brought in a verdict of "Not Guilty," which was received with enthusiasm by the public in court.*

This problem of punctuation may arise directly if a message is sent by telegraph without due care to indicate the position of the original stops. For example, there is the actual case of the so-called

---

* For further details of these cases, see Marjoribanks's *Life of Sir Edward Marshall Hall.*

"Letter of Invitation" sent by the Reform Committee in Johannesburg to Dr. Jameson, across the frontier at Pitsani, before the Raid. As originally punctuated, the crucial passage seems to have read thus: "It is under these circumstances that we feel constrained to call upon you to come to our aid should disturbance arise here. The circumstances are so extreme that we cannot avoid this step, and we cannot but believe that you, and the men under you, will not fail to come to the rescue of people who would be so situated." In this form, the appeal is obviously a conditional one, and Jameson was asked to break into the Transvaal only if a disturbance had already arisen in Johannesburg—an event still in the future. But when this message was cabled to London, apparently no indication of the punctuation was given; and *The Times* was left to guess it. As a result, the full-stop after "aid" was shifted over four words towards the beginning of the message and the result became: "It is under these circumstances that we feel constrained to call upon you to come to our aid. Should disturbance arise here, the circumstances are so extreme that we cannot avoid this step," etc. The change in the position of the full stop converts the original conditional appeal into the equivalent of "Come immediately. If you don't come now and if disturbance arises, things will be in a desperate state."

A similar example is to be found in a case of murder which occurred in Hampshire in 1867. A young man, Baker by name, was arrested on a charge of murdering and dismembering a child in a horrible manner. Baker's lodging was searched after his arrest and his diary was discovered. In it, under the fatal date, was the entry: "Killed a young girl to-day. Fine and hot." The prosecution naturally quoted this as a confession. The defence, however, contended that Baker had omitted a comma, and that the entry should be read: "Killed, a young girl to-day," which might refer merely to Baker having heard from someone about the murder And in support of this reading the defending barrister quoted other entries from the diary in which a necessary comma had been left out, such as: "Played a cricket match to-day." This, as printed here, would imply that Baker took part in the match, but apparently he was merely chronicling the fact that a match had been played; and this would have been plain if a comma had been inserted, thus: "Played, a cricket match to-day."

This ingenious contention failed to convince the jury, who returned a verdict of "Guilty," the correctness of which was shown by the subsequent full confession of Baker. Had the case been tried in our day, there would doubtless have been chatter about schizophrenia and so forth; and the murderer might have been reprieved. But they were less instructed—and certainly less sympathetic—in 1867, and Baker was hanged.

But let us return to the problem of intonation. Anyone who doubts its importance in testimony may usefully reflect on the following example. Suppose we write: "He is a fine artist." The five words apparently express a clear idea. But these five words in this order may be spoken so as to convey various meanings, according to which word we choose for accentuation and the tone we employ. "*He* is a fine artist." Spoken in one tone, that means that the subject is a fine artist, as distinguished from others who are not so gifted. On the other hand, by speaking in a tone of contempt, the sentence becomes a colloquial expression indicating disdain for the subject. Let the reader make the experiment of accentuating each of the five words in turn, and he will find that the phrase can express quite a number of meanings; yet in each case the same five words are being used in the same order. Again, even the monosyllable "Oh" by different intonations may be used to indicate pain, surprise, contempt, doubt, or regret. Instances like these prove beyond dispute that it is not sufficient merely to memorise the words themselves; if the witness is to reproduce the sense of the speaker's remark, he must record the intonation in his mind as well.

We have traced the impressions of a witness up to the point at which they have been recorded in his memory. But now enters a new factor—forgetfulness, and that gradual effacement which is familiar to us all. Attempts have been made to determine experimentally the rate at which data are forgotten. One investigator apparently satisfied himself that on the average a witness forgets per diem 0.33 per cent. of his original impressions; but figures of this kind mean very little, since the memories of different people differ so much from each other, and some facts are more memorable than others. What is quite certain is that oblivion does overtake many of the data which the mind has initially stored up.

In the "Rising Sun" murder case* in 1907, a carman was called as a witness by the prosecution. He had served for twelve years with the Aerated Bread Company and for sixteen years in the Army, which he left with the best possible character. The defending counsel, Marshall Hall, went out of his way to state that he had no intention of attacking this witness; the best man in the world, however, might be an untrustworthy witness. Here, then, in this man MacCowan, the Court had a thoroughly honest man doing his best to give accurate evidence. The whole question at issue is the exactitude of his recollections; and the fairest thing to do is to quote from the shorthand report of the trial at the point where MacCowan was cross-examined by Marshall Hall. The witness had seen a man in the street, and the crucial point was the exact time at which this took place:

> Before the magistrate and the coroner you said the man left No. 29 as nearly as possible at five minutes to five?—Yes, and I stand by that.
> Do you now say that it was twelve minutes to five?—Yes, I walked the distance and found out I was wrong since that morning. Going by my clock at home, I put the time at twelve minutes to five.
> Have you altered that statement because you were cross-examined before the magistrate as to when the street lights were put out?—No, I had not then timed myself in walking the distance.
> Were the standard lights out?—No, I am certain that they were in.

* The quotations used here are taken from the shorthand report of the case, reprinted in *"The Trial of Robert Wood,"* in the series *Notable British Trials*.

> What sort of morning was it?—It was a drizzly, thick, muggy morning.
> You still stick to the "drizzly, thick, muggy" description? —Yes.
> Do you know that not a drop of rain fell in London that day?—I call weather like that when there is dew, "muggy." I have not swallowed the dictionary. I am Suffolk. That is how we talk in Suffolk.
> In Suffolk a morning not raining is drizzly?—Yes.
> What kind of morning was it?—It was a foggy morning.
> You have described it before as drizzly, thick, foggy, muggy morning, getting daylight, but not yet daylight?—Yes.
> Were the standard lights out?—No, I am certain they were in.

As a witness for the defence, Marshall Hall called an official of the Electric Light Department of the St. Pancras Parish Council. This witness explained that the time when the lights were cut off was recorded by an instrument.

> Could the lights possibly have been lit at five minutes to five?—No, it is quite impossible.
> Could they have been alight at 4.48?—No.
> What is the latest the lights could have been on?— 4.40.

Quite clearly, the witness's recollections of seeing the lights were erroneous. Further, as to the "drizzly" morning, the weather records show that that week in London was the hottest of the year, and that no rain fell either on that day or the one before it. The murder occurred on September 12, 1907, and the witness gave his evidence on December 13. Thus, in just three months, he had forgotten the details of two important points: the street lighting and the weather. These were important, because it was a question of identifying a man whom he had seen, and the value of the identification would depend greatly on the available light and the clearness of the atmosphere.

An even more curious case of shortness of memory is exemplified in the following extract from the same trial, dealing with the cross-examination of a witness by Marshall Hall.

> Did you hear that he slept with Dimmock* on the Monday, Tuesday, and Sunday nights?—Yes.
> Did you know that on Wednesday morning Roberts had accompanied Dimmock home?—Yes.
> Did he tell you in the course of a conversation?—Yes.
> Was not that a curious conversation to hold with a man who had only known you, and you him, for three days—to tell you he had passed the three previous nights in that way?—He did not say that. He merely mentioned he had been with her.
> For three nights?—No.
> Why! I put the three nights to you specifically. I know I did, because, getting an answer I had not expected, I put the question again specifically—and put the three nights in the wrong order purposely to mark the question.—I did not understand.
> Mr. Justice Grantham: No, nor did I.

Here, then, within at most a couple of minutes, both witness and judge had forgotten that three nights had actually been specifically named and agreed to by the witness. Yet, in a trial for murder, one would expect that the judge would be on the alert, and would note the forgetfulness of the witness instead of, as in this case, sharing it.

* The murdered girl.

After he has embedded a record in his memory, and after it has lost something there, a witness has to extract what he can from the remaining material before he can give his testimony. And just as his personal idiosyncrasy intervenes during the selection of the material which he stored up, so now his proclivities will to some extent govern the process of excerption. He will not produce all the details which he remembers. Instead, he will bring out only those which he thinks of importance, leaving the rest untouched. Once more the material will suffer a deformation. And here again, if he is testifying in Court, his evidence may suffer because he tends to lay stress on matters which seem to him important, while he neglects others of less interest to him, though they may be of the greatest pertinence in the legal case which is being tried. Further, he may incline to select the more sensational incidents from his memory and pass over those which seem to him less striking.

We now come to the stage where the witness has to put together his memories and make a presentation of them in words. This is much more difficult than it appears at first sight. Suppose, for example, that a man has discovered a bush of white heather somewhere on a trackless moor. He can find his own way back to it. Countless tiny clues will spring into his memory as he goes. But if he tries to explain to a friend how to reach that clump of white heather, he will be in most cases unable to make the friend "see" the path which should be followed. "At this point you will see a stone of a certain shape . . but I can't describe exactly what the shape of it is. Then you'll come to a burn, and you must follow that down until you come to some shingle. I can see the look of that shingle in my mind, but I can't tell you how you can recognise it from other banks of shingle which you'll pass on the way." And so forth. The narrator has a perfect mental picture of each stage in the journey, but he is absolutely incapable of "putting the thing into words."

Now a somewhat similar problem faces a witness who tries to describe a stranger whom he has glimpsed casually. Everyone knows this difficulty from their private experience. "He's got blue eyes, a thin nose, thick lips, florid complexion, thinnish dark hair . . ." and so forth. That might fit 10 per cent. of the population of any city. The police get round this difficulty to some extent by training their observers to note certain details carefully, and so produce what is called the *portrait parlé*. But the ordinary layman has to depend on what he notices casually; and when he attempts to put his impression into words, he fails ingloriously to produce anything of much value for recognition purposes unless his subject is very striking in certain aspects.

Then, again, the vocabulary of the witness will affect the value of his testimony. He may have heard the sound of a blow, but he may not have words with which to distinguish clearly between a hammer-stroke and the concussion on a punch-ball. Or, in describing the clothing of a person, he may not be able to differentiate verbally between "pepper-and-salt" and "herringbone" patterns in cloth. His description of shapes will be limited by his experience or inexperience of comparable objects. For instance, the reader is perfectly familiar with what is called "metallic lustre." But let him try the experiment of finding a description of this which will differentiate it clearly from the lustre on the surface of glass, or the lustre shown by polished mahogany, and he will discover that it is not easy to find words which will serve.

We have now reached the final factor which may influence the testimony of a witness: the crystallization into a rigid form which a narrative is apt to undergo when it has been told several times in succession. It is the unimaginative witness who is most prone to this fault, for the imaginative one is apt to fall into the opposite error, and embroider his tale with extra details which may possibly be inventions unconsciously thrown in to make for vividness. The unimaginative man is inclined to stick more and more closely to what he said before, as this saves him the trouble of forcing his memory; and thus his story falls eventually into a fixed mould. The exact reproduction of the same details time after time certainly gives the impression that the witness is accurate; but what actually happens is that he is remembering the tale he told last time, and not the original impressions which he received; and an oft-repeated story may be inaccurate at its foundation. The mere reproduction of the same details, time and again, is no guarantee of the veridity of the details themselves.

But practice is better than theory, and in conclusion it seems desirable to suggest some examples which will enable the reader to test his own value as a witness by trying to answer a series of simple questions upon subjects which are matters of his everyday experience. To make the test a fair one, he must co-operate by putting his answers into words spoken aloud. Merely to 'think' the answers is not enough, since he must place himself in the position of someone giving oral evidence in the witness box, where a precise verbal expression is demanded.

And now for the problems themselves. What were the size, shape, colour, and pattern of the cup which you used at breakfast this morning? What was the make of the third motor car which you encountered to-day? If you looked at shop windows, name ten articles in the first window you examined. After leaving your house this morning, where did you first encounter a man smoking a pipe? What was the subject of the biggest advertisement on the last hoarding you noticed before reading this? Where were you at 4:15 p.m. a week ago?

Perhaps the reader will say: "I never notice that sort of thing; it doesn't concern me." But if you were put into a witness box it is highly probable that you might be asked about a good many things which would normally be quite outside your range of interest also. However, let me meet you here by putting some questions dealing with things which—if you are an ordinary person—you must have seen and handled many, many times in your life and with which you might therefore be supposed to be familiar: a set of playing cards.

In the nine of spades do the sharp points of the pips all point in the same direction? The king of diamonds has three diamonds in his belt; what has the king of hearts in his belt? In which direction do the various knaves face, and do the corresponding kings face in the same way, suit for suit? How many spades are there in the space under the spade queen's hand? Which kings and knaves carry axes? Which kings have their hair curling inward? Which king has no moustache? When you have answered these various questions and checked the answers by examining a pack of cards, you will have a pretty fair idea of whether you tend to observe a group as a whole or its component parts.

VIII
ELEPHANTS FOR WANT OF TOWNS

THIS EVENING I TOOK DOWN from its shelf R. L. Stevenson's *Wrong Box*. I have re-read it often and hope to do so many times more. Like the *Alice* books of Lewis Carroll, and Boswell's *Life of Johnson*, it can be quoted appositely in almost any circumstances which arise in everyday life; and that quality marks a book out from the common run.

Ardent Stevensonians discourse to me of Stevenson's various villains: Long John Silver, the President of the Suicide Club, John Vandeleur the Dictator, the Master of Ballantrae, and other scoundrels polished and unpolished; but I would gladly sacrifice the lot if thereby I might preserve for my entertainment that amusing and unscrupulous person Michael Finsbury, even though "he looked a trifle too like a wedding guest to be quite a gentleman." Charles II, with his usual directness, complained of one of his relations: "I have tried him drunk and I have tried him sober; and, drunk or sober, there is nothing in him." Charles could hardly have made the same reproach against Michael Finsbury, who, whether drunk or sober, is always diverting. In these days, when the world grows over-complex for the taste of many people, a cry arises for what is called the literature of escape. I have much pleasure in proposing the Finsbury family as leaders of an Escaping Club.

Open *The Wrong Box* at random and we chance upon Morris Finsbury in one of his rare moments of decision, confronting his cousin at the far end of the main line departure platform on that eventful Sunday afternoon, when all around him "the vast and gloomy shed of Waterloo lay, like the temple of a dead religion, silent and deserted."

"But I'll tell you one thing—I'll have that eight hundred pound—I'll have that and go to Swan River—that's mine, anyway, and your friend must have forged to cash it."

So cried poor Morris Finsbury, in righteous indignation, as we may read in the fourteenth chapter of *The Wrong Box*. But where was Swan River, and why should Morris Finsbury select that spot, of all places in the world, as a suitable locality in which to start a new career after his ill-starred leather business had foundered beneath him?

A philatelist will give you the answer to the first question immediately; for he remembers those stamps bearing a graceful swan afloat on calm water, which pleased my fancy so much in earlier days. Younger folk, who are not stamp-collectors, may well have more difficulty over a reply; for, like John Finsbury when he posed as The Great Vance, the Swan River Settlement adopted an alias and became known to a wider public under the new name of Western Australia. So much for the location. But how did Swan River come by its name?

Up to the very end of the seventeenth century, white was the only colour known in swan plumage; and when people spoke of a "black swan" they meant something rarer than a "white elephant"—something of the nature of a "blue moon." By a singular stroke of fortune, we are able to name the precise day when this locution fell out of date. On the 6th of January, 1697, the Dutch navigator Willem de Vlaming, cruising off the coast of what was then known as Zuidland, observed an estuary and despatched a couple of his boats to explore it. To their amazement, the sailors saw first two, and then more black swans. They secured a couple of live specimens of this strange bird, a veritable *lusus naturæ* as it must have seemed to them, and took them to Batavia.

News of this amazing find filtered back to Amsterdam in due course, and came to the ears of the burgomaster of that city, Witsen by name. By a happy chance, Witsen was a Fellow of the Royal Society, and he communicated the tale of the discovery to Martin

Lister, through whose agency the facts were published in the Society's *Philosophical Transactions* in 1698. So Swan River came by its name, and thus the emblematic swan has found its place in the albums of philatelists.

Later explorers, Captain Cook among them, found that the black swan existed in many other parts of Australia and was abundant in some places; but it seems to have decreased rapidly in numbers as a wild bird, and nowadays there are probably more black swans in other lands than there are to be found at large in their mother-country.

So much for the question: Where was Swan River? But to the further inquiry—why Morris Finsbury selected Swan River as a suitable goal for himself and the eight hundred pounds—I can give no answer; for it passes my understanding to know why poor Morris, having made a most complete muddle of his leather business, should suppose himself in any way adapted to make even a livelihood, much less a fortune, on the soil of a country in process of development.

These are mere minutiae, however; bigger changes are recorded in the atlas. In his *Heart of Darkness*, Joseph Conrad painted one of the impressions left by the old maps in their time. "Now when I was a little chap I had a passion for maps. I would look for hours at South America, or Africa, or Australia, and lose myself in all the glories of exploration. At that time there were many blank spaces on the earth, and when I saw one that looked particularly inviting on a map (but they all look that) I would put my finger on it and say, When I grow up I will go there."

Conrad's pastime was mine also, and I remember with especial gratitude an old white-bound volume, part atlas and partly descriptive, over which I used to pore "when I was a little chap." At the time it was printed, Stanley had never set foot on the Dark Continent. Africa appeared in its pages as an outline with harbours sprinkled around the coast and a central tract devoid of place-names: not a lodge anywhere in that vast wilderness. The Prophet Jeremiah and the poet Cowper need have sought no further, if they were really genuine in their expressed cravings for reclusion.

Even when the names strayed a little inland, one had only to consult the text in order to learn that they had been planted amid the wilds. The nature of the Great Karroo, for instance, was elucidated by a quotation from Pringle, which began:

> Afar in the desert I love to ride
> With the silent bush-boy alone by my side. . . .

The compiler of the volume had obviously felt that this vast area of blank paper might suggest that he was not giving full value for the purchaser's money; or he may even have feared that he would not get the purchaser's money at all, unless he bestirred himself. Perhaps he took a hint from one of Swift's poems:

> So geographers, in Afric maps,
> With savage pictures fill their gaps,
> And o'er unhabitable downs
> Place elephants for want of towns.

Certainly, with a lavish hand, he had scattered over the face of Africa not merely portraits of elephants, but of lions, giraffes, ostriches, and crocodiles as well, populating the void and making the desert busy as a market place: for according to the scale of the map his elephants were about 200 miles long, and elbowroom must have been restricted by the time the map was finished.

Alas! Things have grown worse since those days. No longer does the geographer place elephants for want of towns; the towns have crowded the unfortunate elephants clean off the maps. This is hard on the young. What right-thinking boy would wish to delete an elephant, or an ostrich, or a giraffe, merely to see it replaced by some prosaic name like Newcastle, Aberdeen, Harrismith, or Barberton? But the damage is done now. *Où sont les neiges d'antan?* Where are our lions of yesteryear? The elephants have departed from our modern banal cartography and must be sought in their last fastnesses, the *Children's Encyclopaedia* and other works wherein the old tradition is still respected.

But nature abhors a vacuum; *tamen usque recurret*. The Dry-as-dusts, with their chains, staffs, clinometers, and theodolites, have driven Wonder from the atlas, but it has returned by a door through which they can never pass. What the schoolroom has lost, the nursery has gained. Salute, then, Mr. Bernard Sleigh* who has ventured into Fairyland and brought us back a map of that entrancing country; nor must you forget Miss Daphne Miller† who, like a new Herodotus, has noted down for us the manners and customs of its inhabitants.

Fortunate indeed is the child who finds, one day, pinned up on the nursery wall, this pictorial chart of Oberon's kingdom. For it is pictorial. All the nursery favourites appear on it, each *in propria persona*—except Tom Thumb. But even he is not forgotten, for an inscription informs us that "Tom Thumb is somewhere here but he is too small to draw." With a map that is all his own, five feet in length and eighteen inches high, printed in seven colours, no proper inmate of the nursery need lack amusement. He can start at the Mermaids' Rock where he can listen to their songs or, if he has a big shell to hand, he can hear the murmur of the sea as it beats below. Then he can embark on Ulysses' ship or join the Argonauts and so cross over to harbour in Fairyland itself. After that the pointing of a finger will show him the way. Breakfast can be had at Miss Muffet's House, as the Spider does not seem to have followed her home. It is, of course, useless to try pot-luck with Old Mother Hubbard; we know that her cupboard is bare. Then, with a nod to Goosey Goosey Gander and a wave of the hand to Humpty Dumpty on his Wall, the young explorer may proceed to Banbury Cross after turning aside to drop a pious tear on Cock Robin's Grave. Jack Horner could be found at his cottage; but we know what kind of fellow he was. It is much better to skirt the shore of Kelpie Bay and seek out Ole Lukoie, because there is no one in the whole wide world who knows so many tales as he does. After that,

---

* *An Ancient Map of Fairyland*, by Bernard Sleigh.
† *Travels in Fairyland*, by Daphne Miller.

back by Bogies' Corner and Rapunzel's Tower, until we meet Mrs. Bond who directs us to King Thrushbeard's Castle and dinner. The walk finishes with a visit to Hushaby Baby on the Tree-top, a nursery tea with Jack Sprat and his Wife, and so to bed in the House that Jack Built. On another day one may seek King Arthur in Avalon or journey to the Other End of Nowhere and meet Mother Carey. And still there are all the people of the Greek myths to visit, and the folk out of Norse mythology. If we have not heard of them before, Miss Miller will enlighten us.

But we cannot linger for ever in the nursery. Peter Pan begins to "date" sadly; and it is in the essence of his nature that he can leave no posterity to carry on his tradition. Let us abandon our five-year-old adventurer to plan for himself fresh itineraries on his map, whilst we close the nursery door and cross the hall to the library. There we can switch on the reading-lamp, take down from the shelves some volumes of maps and charts, and settle into a comfortable easy chair by the fire, with the hearthrug as our magic carpet.

To those of an imaginative turn, few things are more stimulating than an atlas. Time and space no longer exist for him who holds the proper passport; he may flit from continent to continent or from sea to sea at his ease, without stirring from the circle of light about his reading lamp.

> How many miles to Babylon?
> Three-score miles and ten.
> Can I get there by candle-light?
> Yes, and back again.

And it is a question whether this form of travel, if it be followed in the true spirit, is not more satisfactory than any actual voyaging. Real cities, seen by the mortal eye, seldom attain the grandeur which phantasy has lent them; modern Rome can never be an unshorn embodiment of all those centuries of legend and history which left it where it stands; nor can the deserted temples of Egypt give back to us in the flesh the spirits which once thronged and worshipped within their walls. Gone is the lamp that Hero lit to guide Leander; and our modern lights, efficient though they be, can shed no radiance like that which still shines from afar across the waves of Time.

Many are the influences which sway your true atlas-lover in his journeyings. In some continents, old associations guide him and he loves to dwell for a time amid regions famous in history. There, in the darkness beyond his reading-lamp, he catches glimpses of the beach beneath the walls of Troy; the lines of camp fires sending up their smoke to mingle with the falling dusk; Ulysses at his tent-door, burnishing his weapons and weaving fresh schemes; Achilles in bright armour driving his war chariot along the strand. A turn of the page, and another shore appears, with Leonidas and his Three Hundred watching their last sunset and preparing to confront on the morrow all the hosts of Persia. Or, again, there may swim into view a high garden in Carthage where Hamilcar Barca paces to and fro beneath the trees, pondering bygone voyages or planning vengeance on Rome.

Where the perfume of history clings but faintly, there are other magnets to draw the contemplative traveler onward. Ispahan, Bokhara, Samarkand: something in the heavy syllables leads him forward, deeper and yet deeper into Asia. The places themselves matter not at all; they are mere names round which romances may be woven; but they conjure up visions of strange cities lost in the wilds, of mysterious hordes roaming the desolate lands, and of marts into which the caravans streamed endlessly, bearing all the riches of the East.

If solitary rovings grow wearisome, one can enlist under a trusty leader and follow Marco Polo by Balkh, Kashgar, Yarkand, Lob Nor, and so across the Gobi desert to the court of Kublai Khan. Or, if sea-faring be one's pleasure, one may embark with Erik the Red for Greenland and watch the founding of those ill-starred colonies whose fate is still a mystery. And one may sail with Leif Ericsson to America, visiting Helluland, Markland, and Vinland, and meeting with those ugly dwarfs, the Skraelings. Or one may join company with Magellan on his voyage to the Spice Islands. Or one may enlist under Captain Cook and follow him to his death in the depths of the Pacific.

Then there are other pages of the atlas with charms of their own. The traveler haps upon the Everglades and blesses the poet who christened them. What boots it that the name covers a dismal and impenetrable maze of swamp and woodland, rotting in the heat and moisture of the tropical sun? Here, if ever, the atlas is better than the pack mule; he who wanders in it has leave to dream of vast cool forests, bird-haunted, through which he may stray as he shall please.

And should he tire of these endless arcades, it is but a step to the Bay of Ten Thousand Islands, where he may watch the sea foaming of a morning around these multitudinous beaches. Or he may go further afield and in eastern waters seek out names which seem to have strayed out of some nightmare rather than to have been born of any kindly human speech: Calapan, Bohol, Bislig, and Calbayog. Or he may turn northward and circle the shores of Greenland with its litany of despair: Cape Discord, Cape Desolation and Cape Farewell.

But if the traveler be well advised, he will put aside his modern volumes and take down instead from his shelves some yellow-paged atlas of the seventeenth century; for it is only in works like these that the pure joy of voyaging upon the chart can be attained. The ancient paper, the dog's-eared pages, the peculiar perfume of the well-worn leather binding: all conspire to banish the everyday world and to open the road into lands that never were. The very names have a different sound: "One Eye Savanna," "Graves Valley," "Pirateer Quarters"; for such as these, who would not exchange a hundred Brownvilles in the State of Nebraska, U.S.A.? And does San Domingo bear with it the associations of lost Hispaniola? Truly, with the flow of the years, something has gone out of our maps. Aforetime, with their doubtful renderings, they were finger-posts to the undiscovered lands rather than mere guides to the known; and with the trigonometrical survey, some faint and elusive spirit flitted from the atlas, never to return.

Your true traveler by candle-light cares nothing for exactitude in topography. To him, it is the Unknown that beckons; and the sedulously triangulated map is but a hampering instrument on his journey. In these earlier charts, the Unknown is ever at hand. There may be strivings after accuracy in the marginal drawings of natives entitled: "The Inhabitants are Thus Attired"; but such things hinder little when the mind is free to rove across blank spaces sprinkled with sparse jottings: "Many Villages and Some Castles" or "Great Mines here." There may be a rough, uncertain outline with the confessing legend: "Lake not Well Known"; and immediately before the mind's eye there rises a picture of some forlorn explorer breaking through a belt of woodland and coming suddenly upon the shore of some vast new inland sea. Startled wildfowl rise screaming from the reeds about his feet, and the waves lap upon the beach, untrod till now by any human creature. It is the furthest point on his journey. Longingly he gazes at the wide sweep of the strand; eagerly he scans the waters, seeking for the further shore. But westward he sees nothing save a wooded islet here and there; beyond them, the lake stretches unbroken to the horizon. Regretfully he sketches in the outlines of the nearer beaches; he adds his humble confession of ignorance; and then he turns his back upon this new discovery that he may bring the knowledge of it to his fellow-men. And so it stands in the atlas, honestly told in the fewest of words: "Lake not Well Known."

In this immanence of the Unknown, the traveler by candlelight finds the greatest charms; and for him the intrusion of the exact geographer has few terrors. When even the wilderness of the Amazon has been surveyed, and when on Earth there remains no spot uncharted, he may put his terrestrial atlases back on his shelves with a sigh of regret; but it will be only to take down instead the work of other cartographers. Had the traveler lived in earlier years, when Galileo's telescope brought new worlds within his ken, he might have voyaged to the Moon and mused by the shores of those Seas whose very names bring peace in their syllables: Mare Serenitatis, Mare Tranquilitatis. But the progress of science has robbed him of this haven. Who would seek repose of spirit on these arid plains, sunblasted by day and frost-riven by night? What adventurer would set forth on a cold hunting across the bleak hills of the Moon? All that is left for him is the further face of the satellite, for ever unknown to man; and of this region there are no maps to guide him in his wanderings; his tiny lamp would fail to pierce its mysteries. Let him rather take down another volume and pass to Mars. Here, at least, are warmth and light and changing seasons; and he may meet strange beings by the banks of Tanais or on the shores of the Lake of the Sun.

But even across this gulf of space, science may yet stretch out its hand and destroy the marvelous as it has done before; so that in the end the planets may become familiar to us as our gardens and the ring of encircling Mystery may be pushed a little further back. Where, then, will the traveler by candle-light seek for new windows opening upon the Unknown?

Now he may return his atlases to their shelves. They will help no longer. The candle which lit the way to Babylon may be blown out, for only the light that never was on sea or land will serve him upon his further journeyings. By the aid of its visionary gleam, he can voyage into regions beyond the margins of any chart, discovering the Auroras, Isla Grande, or else the Nimrod Isles, and making his landfalls on their elusive coasts.* He may even, if he be lucky in the time of his expedition, glimpse that enchanted land of joy and beauty, Hy-Brazil—or O'Brazil, for its name seems indeterminate like itself—which appears but once in every seven years on the blue rim of the ocean to the west of the Aran Islands.

But better at times is it for the wanderer to take ship under some trusty master and to set sail for lands that never were. Ulysses will guide him past Cythera on the nine days' voyage to Lotus Land, then onward to the floating island of Aeolus, ringed by its wall of unbroken bronze, and so to Aea, with Circe's fair halls of polished stone amid the forest glades, and yet further between the quicksand and the devil-fish to the hollow caves of Calypso's Isle.

Or, if he so choose, he may go with the wise Pantagruel to the Isles of Tohu and Bohu, and past the confines of the Frozen Sea where he will hear what he will hear, and so to the Ringing Island and the land of the Furred Law-Cats, where he may lose the gold

---

* An account of these may be found in *Oddities*, by Lieutenant-Commander R. T. Gould, R.N.

in his purse. And beyond that he is free to sail to the Isle of Odes, where the highways are alive and the roads go up and down, talking as they go, till at last he comes to Lantern Land and descends the Tedratic Steps to seek counsel with the laconic Oracle of the Dive Bouteille.

If he seeks a more devout pilot, then let him join Saint Brendan. It is related—but Allah alone knoweth all!—that in the sixth century A.D., this holy man constructed a well-ribbed boat covered with oxen hides tanned in oaken bark and smeared at every crevice with good Irish butter. Fourteen monks embarked with him, and they took with them food for forty days and a keg of butter to treat fresh skins. Setting forth on their mission in the name of Father, Son, and Holy Spirit, they left the coast of Ireland and sailed for fifteen days into the northwest, when the wind fell and they betook themselves to the oars. Their store of food was exhausted just as they drew near an almost inaccessible, precipitous isle like Ailsa Craig or St. Kilda.

Landing here, they were welcomed by a dog which led them to a luxurious home, richly prepared for the comfort of the voyagers. Things are not always what they seem. Brendan soon discovered that this was all a temptation devised by the Devil, who, by his spells, struck down one of the company.

Leaving this ill-omened place, they set sail with a favouring breeze and, on the eve of Easter, they reached an island whose many streams swarmed with fish and whose fields were white with sheep* as great as oxen. From an inhabitant, the monks learned that no man milked the ewes and that winter never pinched them, which accounted for their hugeness.

---

* Learned commentators have pointed out that there is a Sheep Isle in Sanda Sound, at the south end of the peninsula of Kintyre, that one of the Orkneys is called Shapinsay, or Sheep Island, and that the name of the Faroe Islands is supposed to be derived from the number of sheep (*faar*) found upon them.

This man explained to the explorers that they should spend Easter on an island beside the Paradise of Birds, and to this they came: a strange, stony land without port or beach or turf. Brendan himself prudently remained in the anchored boat; but the rest of the company went ashore, sang Masses, and lighted a fire to cook their provisions. The flames sprang up, their pot boiled over, whereupon the island rustled, moved, and fled into the wastes of the ocean with the blazing beacon on its back, whilst the terrified monks tumbled back into their fragile craft.* Brendan, rather late in the day, then explained to his panic-stricken companions that this island "was the greatest of sea-monsters. Its name was Jascon, and it amused its leisure by trying to bite its own tail, a feat which it had often failed to achieve owing to the grossness of its body. Jascon, it appears, was a friendly creature, for in the end they kept Easter on his back—"a difficult mode of piety," as an ancient Irish commentator remarks—and he was good enough to transport them to the Paradise of Birds.

Later, they touched at an isle, grassy and full of flowers and trees, and inhabited by snow-white birds which, they learned, were the spirits of the dead. One of these predicted that Brendan would wander on the seas for seven years. The company quitted the isle, leaving the birds singing the praises of the Deity.

The explorers later touched at the Isle of the Family of Ailbe, and spent Christmas on it among the Silent Monks who had lived there since the time of Saint Patrick and Saint Ailbe: a weird company who never spoke, nor aged, nor changed, and who cooked no earthly food but relied on God to nourish them.

Thereafter Brendan and his comrades penetrated into the far north, encountering many strange things, among them Judas, on an iceberg, on which he was permitted to cool himself on feast days.

Finally, Jascon piloted them back to the Island of Birds, and forty days of further voyaging brought them in darkness to the

---

* The reader will recall a similar incident in the First Voyage of Sindbad the Sailor.

Promised Land. When light dawned, they wandered amid its fragrant orchards and by its mighty river, gathering souvenirs in the form of fruits and gems. Forty days passed without a night, then they set sail again and returned to Ireland.

If these wonders will not suffice, the traveler by candle-light may change the face of the globe for his pleasure and may bring above the waves those lands which vanished in days of yore. He may venture into lost Lyonnesse; he may bring Atlantis from the depths; or, if it be his fancy, he may solve the mystery of Easter Island and its enigmatic statues by calling to the surface the vanished Continent of Lemuria.

But now we have traveled further than Babylon, and our candles are flickering to their end. It is time to replace our atlases on their shelves and put off further exploration until the mood comes upon us again.

IX
THE MANICHEE

LET US BORROW THE STYLE of the newspaper puzzle column.

"Which companion of Joan of Arc was a millionaire, a mystic, a marshal, a Manichee, a magician, and a mass-murderer?"

Born in 1404, in the castle of Machecoul, on the borders of Brittany and Anjou, Gilles de Rais was connected with some of the noblest families in France: through his father, with the ducal Montmorencies, through his mother with the house of Craon, and with the de Thouars by his marriage. He was heir to the barony of Rais, the senior barony of Brittany, and the wealth of his far-flung estates was, for those days, enormous.

His father died when Gilles was eleven years of age; his mother remarried almost immediately; and the boy was placed under the guardianship of his maternal grandfather, Jean de Craon, Lord of Champtocé and of La Suze, an aged gentleman ("homme viel et ancien et de moult grand age") who, by Gilles' own account, exercised not the slightest influence over him, but left him to follow his own courses. Men came young to maturity in those days. Jean de Craon—perhaps eager to rid himself of irksome responsibilities—looked about him for a match suitable to his ward; and at the age of thirteen Gilles was betrothed to Jeanne Peynel, daughter of the Lord of Hambuie and Briquebec. But Jeanne Peynel died before a marriage had been made; and again Jean de Craon cast about for a desirable bride. He seemed to have found an excellent choice in Béatrix, daughter of Alain IX, Viscount de Rohan, and granddaughter of Duke Jean IV of Brittany; but again death intervened before the marriage.

At this period, the state of France was pitiful. The king, Charles VI, had become insane. The country was rent by the furious quarrels between two powerful factions: the feudally-minded Armagnacs (Orleanists) and the Dukes of Burgundy, who relied on Paris for help, and who had gained possession of the mad king. Brittany was torn by the rivalry of the Montforts and Penthièvres. The effects of the Black Death had not passed away. And over this distracted land loomed the foreign peril from across the sea. We are apt to see the English effort of that time through the eyes of Shakespeare; most of us learned our *Henry V* at school, and absorbed a romantic idea of those times—"Once more unto the breach, dear friends, once more," and similar fine rhetoric has given us, perhaps, a slightly distorted view. The French, who served as the vile body for Henry's experiments, look on the matter rather differently. They see their country "torn and bleeding, sucked dry to the marrow by England, which, like that fabulous monster, the Kraken, rises out of the sea and from across the Straits casts over Brittany, Normandy, part of Picardy, the Ile de France, all the North and down towards Orleans, its tentacles whose suckers, when they let go, leave behind them only wasted cities and a dead countryside."* Agincourt had been fought in 1415; the Burgundians had made common cause with the English; and, by the Treaty of Troyes, Henry V became regent of France, obtained the French king's daughter for a wife, and was to succeed to the French throne on the death of the mad Charles VI.

Into the struggle in Brittany came the sixteen-year-old Gilles de Rais at the head of his own levies and the vassals of his grandfather. His gallantry, his fine appearance, and the strength of the force under his command, drew attention to him and attracted about him a crowd of flatterers who profited by his splendour and prodigality. He came out of the struggle richer even than before; and by his marriage, in 1420, with Catherine de Thouars, he allied himself with a wealthy and powerful family. Then, until his twenty-first year, he disappeared from history. We know only that he took

---

*Huysmans, *Là-Bas*, p. 61.

over the control of his huge estates, fell into the hands of parasites and bad counselors, and acquired habits of pleasure and prodigality.

Meanwhile, the condition of France was passing from bad to worse. The great once-fertile tract between the Somme and the Loire was becoming almost a desert, haunted by wolf-packs and with robbers and free-lances as its only travelers. Tillage failed; and the peasants, getting arms where they could, turned to banditry for a living. In the contemporary literature, mystery plays and moralities bulked largely; whilst on the walls of churchyards the Dance of Death was a fitting expression of the rigours and uncertainties of the time.

The mad king died in 1422, and was succeeded by his son, Charles VII, but the change in rulers worked little immediate improvement in the fabric of France. Selfish, indolent, weary of his vain efforts to raise money, Charles retired to Chinon, where he set up his wretched little court, and in that nest of intrigues, devoted himself to a sordid career of debauch among his sycophants The limits of his effective power were indicated by the contemptuous nickname: "King of Bourges." Two factors were lacking for the prosecution of effective war: funds and faith.

It was at this juncture that Gilles de Rais appeared at the court of Charles, where a man of his type—handsome, intelligent, ready-witted and, above all, wealthy—was sure of a welcome. But he did not linger in Chinon. In Brittany, forces were gathering against the English, and Gilles raised seven companies at his own expense and plunged into the campaign. Some commentators have found ground for disparagement in one incident. After the capture of Rainefort, Gilles found among the garrison sundry French renegades, whom he hanged out of hand, whilst sparing the English; and in this episode the critics seek to trace an early example of those cruelties which marked the later career of Rais. This seems to go too far. In our own day we have seen how such "collaborators" are regarded and what fate they meet when they fall into the hands of enraged compatriots. There seems no ground for ascribing exceptional cruelty to Gilles since he acted just as present-day Frenchmen have done in similar circumstances. According to the chroniclers of the time, he had the reputation of a "bon et hardy capitaine."

But wealth alone is no guarantee of success in war. The later campaign went ill; the English pushed to the Loire and laid siege to Orleans; the king was contemplating a retreat into the south. But at this critical moment the Duke of Burgundy withdrew from the English camp, and the opposing forces became more even. And now the second desideratum was granted to Charles VII. Joan of Arc appeared, burning with fervent faith in victory and claiming to have the guidance of invisible Powers.

At this stage in his career, Giles seems to have closely associated with Joan in the operations which led to the relief of Orleans. He was with her in some of the heaviest fighting, and he rendered fine service at Orleans, Jargeau, Paty, and Beaugency. Nor was the influence of his wealth negligible, for Charles VII granted him a thousand livres "to recompense him for the great expenditure he had incurred" in assembling at his own expense a large body of men-at-arms and bowmen for the royal service. Nor was this all. He was one of the commanders of the escort to the king and Joan in the march on Reims; he was created a Marshal of France; and he and his brother marshal headed the escort which brought the holy oil in the Sainte Ampoule to Charles' coronation. He was with Joan when she was wounded in the attempt on Paris.

But the greatest honour of all was still to come. Charles VII had conferred on Joan of Arc and her family the right to bear the Royal arms; now a similar privilege was to be bestowed upon Gilles. Letters-patent were drawn up, which granted to him the right to add the arms of France as a border to the Shield of Rais "in recognition of his glorious services and to perpetuate the memory thereof." It is true that these letters-patent were never sealed and registered; the mere fact that the documents were prepared is proof enough of the high esteem in which Giles was held at that stage in his career.

Thus, at the age of twenty-five, Gilles was a Marshal of France, a figure at Court who had won the gratitude of his king, a commander of merit, and a companion of that preternatural and enigmatic personality, Joan of Arc.

What were the effects of this association with so practical a mystic as Joan? Effects there must have been. A man of five-and-twenty could hardly be untouched by an influence which impressed that middle-aged, hard-bitten and energetic rascal, La Hire. Gilles had been an eyewitness of a series of events which amounted to little less than a miracle; and these were credulous times, when the material world seemed but a screen behind which real but invisible Powers were locked in furious contest. Without improbability, one may well see in Joan the origin of that mysticism which was to play so terrible a part in the later life of Giles.

After the death of the Maid, only sporadic traces of Gilles' military exploits occur in literature; and we pass on to the period when he quitted the Court, retired to his Brittany fastnesses, and began that spendthrift career which, in less than eight years, dissipated the greater part of his vast fortune.

His income approximated to £100,000 a year, but he seems to have expended half as much again as a normal practice. As a Marshal of France, he felt it incumbent to maintain a private bodyguard and over two hundred horse. He had knights, squires, and pages to form a little Court such as only the highest in the land could support. He had mansions in Angers and Nantes, and one at least surpassed in magnificence the palace of the Duke of Brittany. His library contained many rare volumes: a strange sight in days when most nobles lacked the elements of book-learning. The theatre was one of his passions, and he celebrated the great festivals of the year by staging elaborate mystery and morality plays. He kept a company of comedians, troubadours and jugglers for the entertainment of himself and his friends; and there were even mauresque dancers, specially procured from Spain.

Nor was religion forgotten. He maintained a whole ecclesiastical household in his castles: twenty-five or thirty clerics living splendidly at his expense and accompanying him on his journeys. He had a cultivated taste in music and was prepared to spend without regard for cost if he could secure a fine voice for his chantry. One of these was induced to enter Giles' service by the gift of an estate, called La Rivière, plus a gift to the man's parents. Mass was celebrated with every magnificence. The vestments were of cloth of gold or of the finest silk; and as it was known that Giles never condescended to bargain over his purchases, the merchants charged him twice or thrice the actual value of their goods. Once he paid the equivalent of £25,000 for three copes of cloth of gold, the actual value of which was probably less than a third of this figure. A goldsmith was procured from Paris, and took up residence. The crosses, candlesticks, censers, and chalices were all of the finest gold or silver, massive and elaborately chased, adorned with brilliant enamels and precious stones. In 1434, he caused two notaries of Orleans to draw up a deed in which provision was made for a projected Foundation in memory of the Holy Innocents, which was to contain a curate, dean, archdeacon, treasurer, canons, chapter and college in perpetuity, with revenues proper for their maintenance; and as trustees he appointed the King of Sicily, the Duke of Anjou and the Duke of Brittany, with injunctions that they should brook no interference with this trust from Gilles' wife, daughter, or any other relatives, friends or claimants.

This for the saving of his soul. On earth, however, he was living—as his heirs complained—"not according to the condition of a baron, but according to that of a Prince." He kept open house to all comers, and guests flocked from all quarters of France to this hospitable roof where they found lavished on them all that luxury and good fellowship could provide.

Gilles was no Trimalchio. Not only did he appreciate expertly the work of others, but he was himself practised in the arts. He spoke Latin with elegance; he could illuminate manuscripts; he could prepare the enamels with which he enriched the bindings of his books; he took part in musical services; and he certainly initiated (and possibly helped to compose) the mystery plays in which he delighted. Such a sweep of accomplishments was prodigious in that rude age, all the more since Gilles had led no cloistered life, but was, in experience, an old campaigner.

This wild craving for glitter and display was soon to bring catastrophe. Huge though his income was, Gilles began to anticipate it; and thus he launched himself upon that slippery slope which led, as it was bound to do, into bankruptcy and disaster. If he lacked the wherewithal to gratify a momentary whim, he borrowed far and wide, pledging himself to exorbitant interest; but no sooner was the ready money in his hands than he squandered it on some fresh caprice and borrowed again to replace it. His parasites reaped a rich harvest from his generosity. He seems to have trusted them completely with the management of many affairs, and they took advantage of this to enrich themselves. In the later stages of his career he was even driven to borrowing from these blood-suckers some of the money which they had taken from his own pocket.

In the furnace of such a monstrous extravagance, even the greatest fortune melts swiftly away; and now borrowing would not suffice. Gilles began to part with his estates.

He sold the domains of Confolens, Chabane, Chateaumorant, and Lombert to a captain, Gautier de Brussac. The estates and castles of Blaison and Chemillé fell to the Lord of Martigné-Briand for five thousand crowns, of which only half seems to have been paid over. Fontaine-Milon and Grattecuisse passed into the hands of the Bishop of Angers. These are mere examples; even the fullest biography of Gilles does not profess to catalogue more than a fraction of the sales, so numerous were they.

But even these sacrifices gained merely some breathing-spaces in that colossal career of extravagance. Gilles continued to buy on credit and to pay dearly for the privilege. Like certain Indian princes, he would set his mind on some object, buy it, and tire of it in a week. Then he would re-sell it for a derisory price or else make a gift of it to some retainer. So little interest did he take in his monetary affairs that he handed to some of his parasites blank forms, duly signed and sealed, which they might fill up as they chose—thus surpassing even the folly of giving a blank cheque.

As time went on things grew even worse. Gilles was reduced to pawning some of his most well-beloved possessions. Favourite books from his treasured library were given in pledge to creditors. Even the vestments and implements of the Mass were hypothecated. Things reached such a pitch that at times Gilles lacked a meal owing to the ill-management of his parasites, though they themselves took care to live "lavishly, like lords of the highest degree, at the cost of the said Messire Gilles."

His total lack of sense in money matters came to be common property at last. In the lawsuits which followed his death, his heirs pleaded that Duke Jean of Brittany, Jean de Malestroit the Bishop of Nantes, his Chancellor, his Treasurer, his other officers, and the nobles, burgesses, hinds, and inhabitants of the Duchy of Brittany publicly held and accounted Gilles to be mad and senseless; and every time they saw him, they mocked and laughed together as at the sight of a fool. They repeated many times and in many places that he was mad, void of sense, and a prodigal. And each of them "sought to acquire whatever they desired of his belongings, knowing that they could make him accept and pass all such contracts as they might please."

Despite interested attempts to shroud such transactions in secrecy, Gilles' family at last learned what was going on; and they stepped in to avert the complete ruin of the Rais inheritance. His wife Catherine, his brother René de La Suze, and the senior representative of the family joined in petitioning the king; and in 1436 Charles VII issued letters patent which recited that, in view of the bad rule and management of Gilles, he was forbidden to sell or alienate any of his lands, lordships, and revenues. Letters were also addressed to the Court of Parliament so that Gilles might be prohibited from selling, and others from buying, any of his property.

This edict precipitated the catastrophe. Duke Jean V of Brittany was both crafty and avaricious; and now he saw his chance. He and the Bishop of Nantes—Jean de Malestroit—refused to permit the publication of the edict in Brittany. Jean V created Giles Lieutenant-General of the Duchy and entered into a solemn pact of brotherhood with him—a *fraternité d'armes* such as had linked Bertrand du Guesclin with Olivier de Clisson. By such means, he persuaded Gilles that he was a firm friend. But the Lavals and Gilles' brother, René de La Suze, had also to be reassured; so Duke Jean V swore solemnly in church, upon the body of the Redeemer, that never, in any circumstances, would he purchase the two estates of Champtocé and Ingrandes. Finally, whilst forbidding the publication of the king's edict in Brittany, he took care that the contents of it were allowed to leak out unofficially. As a consequence, no one dared make an offer for any of Gilles' property lest they should thus bring down upon themselves the wrath of both King and Duke; and in this way Jean V, who cared nothing for the Royal anger, was left in the field as the only possible purchaser—at his own price. As for his solemn oath, its value was proved a year later, since, in 1438, he bought both Champtocé and Ingrandes.

The action of his family had set a term to Gilles' career of extravagance; and this probably accounts for the fact that never afterwards did he trouble to associate with either his wife or his daughter. They retired to a castle, Pouzauges, belonging to Catherine's family. And now Gilles was wholly alone, without a restraining influence, abandoned to the company of tempters and parasites of the vilest types.

Fortunatus's purse was empty at last. But a man like Gilles could not content himself with the prospect of permanent retrenchment. If he could no longer draw from the rent-rolls of his patrimony, then he would seek gold from other sources; and so began the third phase of his chequered career.

The thirteenth century saw alchemy at its high tide. It had not yet merged into the later iatro-chemistry which concerned itself with the actions of chemicals as medicines; and it was still concentrated to a great extent upon three great aims: The Alkahest or Universal Solvent; the Elixir of Long Life; and the Philosopher's Stone which could change base metals into gold or silver. Albertus Magnus, Roger Bacon, and Raymond Lully were great names indeed. Even St. Thomas Aquinas dabbled in the theory of alchemy; and his writings show that he knew the true searcher from the quack. For there are other ways of getting gold than by transmutation; and a horde of sharpers arose who had nothing in common with the true alchemist but sought only to fill their pockets and secure a livelihood by practising upon the credulity of some rich dupe, promising vast and profitable discoveries in return for subsidies. Pope John XXII was so concerned by the growth of this horde of harpies that he issued his Bull *Spondent pariter quas non exhibent*, thundering against all those searchers for gold who promised more than they could perform."

But even the fraudulent alchemist was not without support in high places. He found protection at the court of Henry VI of England; and in France, Charles VII employed one Le Cor, whose services are said to have resulted in debasing the coinage by the minting of quantities of "artificial gold."

Deception was made the easier by the extraordinary yet beautiful symbolism employed by the alchemists,

> "Whose noble practise cloth them teach
> To vaile their secrets wyth mystic speach,"

as we may read in *The Hunting of the Green Lyon*. The stages of their processes were shrouded under names unknown to the uninitiate: Xanthosis, Leukosis, Melanosis. They handled materials disguised by titles like Mercurios Philosophorum and Aurum Potabile. When they wrote books to record their discoveries they chose a cloudy nomenclature calculated to bemuse those outside the pale: *The Triumphal Car of Antimony; The Recovery of the Great Stone of the Ancient Sages;* or *The Revelation of the Hidden Key*. And, as if these were not enough to guard their secrets, they employed also mystical drawings of the Dragon, the Green Lion, the Red Lion, the Lily, the White Swan, and so forth, each of which had its esoteric meaning, known only to the elect. When such things were the common currency of the true alchemist in search of knowledge, any pseudo-philosophical jargon would pass for Gospel on the lips of a quack.

If a patron, weary of mere verbal subtleties, demanded some more practical proof of success, the fraudulent alchemist was ready to supply this. Into vessels containing only samples of the baser metals, innocent-looking powders could be cast as part of the "process"; and these powders contained compounds of gold or silver which, in the final stage of the operation, could be isolated and shown to the dupe as a proof that transmutation had been effected.

Again, pure lead could be melted in a crucible and stirred with an iron rod while mysterious powders were added; and in the end the lead would be found to be mingled with gold. To the gullible, the iron rod seemed solid; but, in practice, it was a tube holding pellets of gold embedded in wax, which melted in the heat and allowed the nuggets to slide unseen into the lead. Or, again, the patron could be shown a long iron nail which, after being partly immersed in some secret liquid, was treated with acid, and was then found to be of pure gold in the submerged portion and of iron in the part which had not been touched by the secret agent. Of course, the "iron" nail had been prefabricated from an iron head joined to a golden point, and the colour of the gold had been disguised by a coating of mercury, which was removed by the acid in the final operation, revealing the yellow tint of the gold. But, as the alchemist was careful to explain, these were merely small-scale results; and to ensure success on a magnificent scale, further investigation—and more money—would be required.

It is not known when Gilles de Rais first dabbled in alchemy. By his own account, his interest in the subject was aroused by a book found in the possession of a soldier imprisoned for heresy in the castle of Angers. This valuable volume—which dealt not only with alchemy but with the raising of devils as well—was read by the Marshal out of curiosity and then returned to its owner. Gilles—apart from his total lack of that curious quality, the "money-sense"—seems to have been a man of wide interests, and it would be likely enough that he should find his fancy caught by the alchemical lure even in the days of his prosperity. These early dabblings in the Great Art were dictated purely by curiosity, and were probably mere dilettantism, with no great driving-force behind them, since Gilles was at that time able to raise all the money which he required.

But after the intervention of his heirs and the consequent financial catastrophe things took a different turn. Barred off from all his older sources of revenue, Gilles found himself at bay. Gold he must have; not on its own account—for Gilles had in him no trace of the miser—but for the sake of things which it would buy: all that pomp and pageantry which had been his very life and which had now been snatched beyond his grasp. The vision of alchemy drifted back within his ken, fascinated him, made him its thrall. The transmutation of the base metals into gold would restore him to riches. But the imaginings of that monstrous intellect far outran dreams of mere wealth. He demanded from alchemy something even more transcendent: "the marvellous power of being able to cast down, as his fancy might dictate, the fortresses and cities which are the best defended by art and nature, without any ever being able to prevail against him." Miracles? But why should he doubt that miracles could come to pass? Had he not stood by the side of Joan of Arc and seen the miraculous in action?

So a laboratory was fitted up in the castle of Tiffauges, and we find Gilles bending over his crucibles and alembics, striving to tear from Nature her inmost secrets. But not alone, for there flocked about him a new horde of sharps and tricksters so numerous that in later days he could not recall even their names. His chaplain, Blanchet, acted as his recruiting-agent, seeking far and wide for adepts in the hermetic art. There were allies, too, among his retainers: his cousin, Gilles de Sillé; his major-domo, Roger de Bricqueville; his reader and chamberlain, Henri de Griart; and his trumpeter, Du Mesnil.

Among his stores of gallipots and crucibles, athanors and pelicans, Gilles untiringly pursued his enterprise. Experiment followed fruitless experiment, recipe after recipe was tested in vain, but ever the elusive mirage lured him on to further toils. Days there were when a gleam of hope deluded the spagyrist; days of doubt when some great trial culminated in ill-success; days, too, when a delusive triumph transported him with joy and lent him the illusion that his long-sought quarry was within his reach. Then, braced by this cordial, he could go forward with renewed hope. But always, at the end of every promising path, there stood failure and frustration, inexplicable; inevitable.

At last he came to realise that the Sphinx refused to yield its secret.

Science had failed him. God, too, in his day, had proved a broken reed. After all Gilles' lavish expenditure on priests, chantries, splendid vestments, golden vessels for the services, costly and elaborate Masses, after even that grand project for the Foundation of the Holy Innocents, the Deity had looked on with stony eyes while the ruin of the Marshal had been consummated. No all-powerful hand had been stretched out to succour him and ward off his doom. So be it! But there was yet Another to whom he might turn, perhaps with better success. If God had no care for him, he would invoke the ancient Adversary and try his luck with Satan. And so began yet another phase in that astonishing career.

In the twentieth century, the flames of Hell have grown dim, and belief in a personal Devil has lost its vividness in many minds.* Still less credence is offered to the idea of a Devil incarnate walking the earth in a garment of flesh and seeking victims as he goes. Thus it is difficult indeed for us to picture to ourselves what must have passed through the mind of a man of the Middle Ages who had resolved to conjure up the Prince of Darkness in material form. What would he expect? One would need a vast and sombre imagination to recapture all the tremors, the horror, the dreadful fascination, the instinctive repulsion, and the frightful apprehension which must have mingled with the exultation of the neophyte as he faced his ordeal. Even the form in which the Evil One would appear was a matter for conjecture, since he might choose any shape that pleased him for the moment. The suave and ironical

---

* It is only fair, however, to quote an instance against this view. After a thoughtful argument between a Protestant clergyman and a Roman Catholic priest, the Protestant was moved to say: "Really, Father, there seems very little difference between our standpoints in this matter." To which the austere answer was: "You think so now; but you won't be dead five minutes before you'll know that there's a Hell of a difference."

Mephistopheles lay in the far future; something much more grisly was in fashion.

Gilles' first efforts in demonomy were merely grotesque. He enlisted the services of an expert, Jean de la Rivière, from Poitiers. The initial step was the concoction of a note to Satan, written by La Rivière in a special ink and signed by Gilles in his own blood. The actual conjuration of the Fiend took place near Machecoul, in a field adjoining an inn and a cottage in which dwelt a wench known as La Picarde. Gilles, his valet, Corillaut, La Rivière, and a fourth man were present. A magic circle was drawn; and when, by chance, La Rivière stepped out of its protection, he was—by his own account—set upon by imps and severely mauled. Gilles himself saw nothing; but, nevertheless, the strain upon his nerves was so great that he was drenched in cold sweat "as if he had fallen into a river, and yet it was not then raining."

A second attempt was made later, but this time La Rivière went alone to the spot, whence he returned with a tale that he had suffered severe mishandling. He had, however, been watched by the woman La Picarde from her cottage; and she had no scruple in declaring that his injuries were only feigned, and that "the most cunning of all the devils were certainly in his own skin."

A third trial was made, La Rivière this time wearing armour and carrying a sword. Along with Corillaut, Blanchet, Griart, and possibly Gilles himself, he repaired to a lonely wood, leaving his companions on its verge. Shrieks, the noise of a struggle and the clash of steel appalled the escort, and the wizard rushed out of the wood, exhausted, trembling, gasping and haggard. The Devil, he explained, had appeared in the form of a leopard, which, to his astonishment, had come straight at him and passed him by without looking at him. He had tried to arrest the apparition, and he showed Gilles some hairs which he had plucked from its neck. These hairs he afterwards burned in the presence of the Marshal to the accompaniment of incantations: an ingenious method of getting rid of evidence. One has a suspicion that some local goat may have been the poorer for part of its pelt; in which case it would be desirable to destroy the hairs lest some inquisitive person should examine them and identify the source.

The unsuspicious Gilles was delighted with the results obtained, and the party spent the rest of the night in a carouse, during which La Rivière extracted from Gilles twenty gold crowns to be expended upon magical appliances for the next attempt. But this was never made, for La Rivière departed in search of the required materials, and neither he nor the twenty crowns were seen again.

He was soon replaced. Wizard followed wizard in Gilles's service, each received with hope but eventually discarded for lack of success. At times Gilles was cheered—and considerably frightened—by what looked like positive results. One necromancer, for example, attempted to call up the Devil in the castle of Tiffauges. A magic circle was traced on the floor of a room and the sorcerer entered it along with Gilles, whilst de Sillé remained outside, seated on the ledge of an open window and clasping an image of the Blessed Virgin to his breast for safety. The magician forbade Gilles to make the sign of the Cross, and then began his enchantments. In the words of Gilles' own account: "I heard voices which were not human, and I became marvellously afraid, feeling that I had confessed myself badly that morning." He prayed to the Virgin, whereupon the wizard hustled him out of the circle. Gilles departed, making the sign of the Cross and slamming the door behind him; whilst de Sillé, also having had enough, fled through the open window. The two forgathered outside the door, when they heard sounds as if someone were beating a feather bed. Despite their temporary panic, Gilles and his cousin were brave men, for, drawing their daggers, they opened the door of the room and found the conjurer lying on the floor, moaning, badly hurt in the face, with a large bump on his forehead. Gilles feared the man might die "whereupon I did have him well confessed by my chaplain, but he did not die."

On the surface, Gilles' proceedings seem to be mutually discordant. To go straight from the confessional to the raising of Satan appears to be a case of "having it both ways" with a vengeance. But further information is available which throws light upon the matter and elucidates the character of Gilles. Hitherto we have seen him in the guise of a "bon et hardy capitaine" afflicted with a ruinous generosity and a mania for extravagance: a trustful and not unlikeable fool. But there was a darker streak in him which grew broader as the years went past, until it blackened his whole nature and turned him into something less than human and worse than a beast.

It will be remembered that, at the outset of his demonomy, Giles gave to La Rivière a "cedula" signed in his own blood. The terms of this document did not transpire until much later, but they contain, probably, the key to the problem. In this "cedula" Gilles had demanded from Satan three things: science, power, and wealth. He had made two express reservations: his soul and his life. So much Gilles himself confessed. But a contract is void unless it mentions a consideration to be paid in return for benefits received; and the Devil might be lawyer enough to know this. What, then, was the consideration which Gilles conveniently forgot in his confession? The valet, Corillaut, was able to remember. Gilles, in return for the three great gifts, promised to sacrifice five little children and give their hearts to the Devil.

This throws light upon the state of affairs as seen by Gilles himself. In his "cedula" he had reserved his soul from the power of the Evil One; he had not yielded himself to an inevitable damnation. But he would still have to account to the Deity for the shedding of innocent blood. Here, however, he was upheld by a belief so strong that it found utterance in some of the last words spoken by him. "There is no sin, however great it be, but God in his kindness and fatherly benignity will pardon it, provided that pardon be asked of Him with great sorrow and contrition in one's heart." The link joining him to the Deity was not broken by his contract with Satan. And so it was possible for him, with his peculiar logic, to traffic with the Devil whilst maintaining—as he did even after his financial ruin—the costly services, the great retinue of clergy, and all the ecclesiastical pomp with which he hoped to propitiate the Almighty. There was always time for repentance. But for a full understanding of that contorted intellect one must turn to the frantic boast which he made during his trial: "I was born under such a star that no man in the world has ever done, or could ever do, that which I have done." It almost equals the pride of Lucifer, and it voices the sincerest belief of the Marshal.

It seems impossible to determine the date at which Gilles entered upon the murderous phase of his career. Vizetelly,* after careful examination of the evidence, seems inclined to put it as early as the lifetime of Gilles' grandfather, Jean de Craon; and he suggests that when the Marshal instituted the Foundation of the Holy Innocents he was endeavouring to atone for child-murders which he had already committed. In his evidence at Gilles' trial, the valet, Corillaut, averred that some of the crimes dated as far back as 1426. Griart, the chamberlain, said much the same thing. But in both these cases there was an evident and comprehensible desire on the part of the witness to name a date which preceded his own entry into Giles' service, so that the evidence is hardly reliable. Whatever be the date of the initial crime, there is no doubt about the sequel.

Brittany has ever been a land of legends. It is the country of the White Wolf, that gigantic beast which hunters have seen but have never slain; and even to-day the peasant grazes a sheep along with his cattle so that if the White Wolf comes he may take the sheep—"the White Wolf's share"—and spare the kine. The immense stony avenues of Carnac stand intact to call up visions of the unknown past and to remind the passer-by of a religion older than history. One may still look down on the walls of submerged Ys, overwhelmed by the sea as Sodom and Gomorrah were by fire. Under Cape Raz lies a dreary arc of sands, the Baie des Trépassés, where the currents beach the body of every sailor drowned on that sinister coast. And there, too, lies that interior sea, the Morbihan, bearing hundreds of islets studded with druidical monuments, mysterious and haunted. Brittany is the home of the enchantress Koridwen, the concoctress of the Miraculous Water, and the korrigans dance by night across its desolate lands. No more fitting

---

* E. A. Vizetelly, *Bluebeard*, pp. 202 ff., 262, 266 (1902).

scene could have been chosen for the vast and gruesome tragedy which marked the final years of Gilles de Rais.

From this haunted countryside began a steady disappearance of children. In a land where superstition had deep roots, an explanation for the earlier vanishings was easy to find: the korrigans, the gnomes had stolen those little victims who were never seen again. Harder heads remembered swift streams, deep pools, the wolves and boars in the woods. Easy for a child to come by mishap if it ventured into dangerous places. But the terror grew. Home after home was in mourning for a little one, strayed and lost for ever. In the end, the horror spread until it embraced the whole region from the Morbihan down into Vendée, and inland as far as Angers. Children would be seen by a neighbour at a street corner and then, when sought for, they had melted into thin air.

Rumours began to pass from lip to lip, all pointing in the same direction; nor were they without some solid support. Gilles de Sillé recruited pages for the Marshal; and when the parents inquired about their boys they were not forthcoming. In one case, de Sillé admitted that the pages were no longer in his master's service; they had been handed over to the English as part of a ransom for the Marshal's brother. A hag, "La Meffraye," was suspected. She was seen talking to a child and that child vanished, kidnapped by some accomplices when the woman had beguiled it into a secluded spot. And so the ghastly business went on, spreading heartbreak and horror into home after home, up and down the stricken countryside.

Why was there no open complaint? To whom could the bereaved complain? In those days there was nothing corresponding to our modern police. Besides, where was the evidence? The children had gone, but no one had seen them go. And over all loomed the power of the Marshal; his hand would be heavy indeed on anyone who dared to raise his voice in complaint, for in those days the peasant had few rights. Nor had the retainer many. At least sixteen of Gilles' underlings were in his secrets, but not one of them betrayed him, for fear ruled them all. They knew what Gilles was, and they trembled. One page who penetrated the mystery was straightway drowned in the moat. Griart, the chamberlain, was so sickened by horror that he thought of committing suicide, but was restrained by the reflection that he would die unshriven; better to live on even amid such scenes, than pass straight into eternal torment. So silence reigned and the Marshal pursued his career, caring nothing for the rumours which ran across the expanse of his estates and even reached his own ears.

Gossip began to focus itself on three ominous names: those of the castles of Champtocé, Tiffauges, and Machecoul. Champtocé lay upon the Loire, about 40 miles upstream from Nantes; Tiffauges stood half-way between Nantes and Cholet; whilst Machecoul lay on the river Falleron in the hinterland of the Baie de Bourgneuf.

It will be recalled that Duke Jean V of Brittany, in defiance of the royal edict, had purchased the castle of Champtocé, but no public transfer had been made, and Gilles still used it from time to time. This was the place where the Marshal's confidence suffered its first shock. Warned of his transactions with the Duke, Gilles' relations decided to frustrate them, and by a *coup de main* to prevent the alienation of the family property. The Marshal's brother, René de La Suze, and his cousin the Admiral, André de Laval, laid their plans for a forcible occupation of both Champtocé and Machecoul, but Gilles in some way seems to have got wind of their scheme. He had to act at once, for there was evidence of the worst in both fortresses.

To Machecoul he despatched his confederate de Sillé with one retainer, ordering them to destroy all traces of his crimes. It was a gruesome task indeed. From the cellars under one of the castle towers they removed the rotting remains of at least 40 children, which they burnt, throwing the ashes into the moat and the river. When, not long after, René de La Suze seized the fortress, the evidence had been obliterated, and Gilles was safe, so far as that plague-spot was concerned.

The Marshal, meanwhile, lingered at Tiffauges. He learned of his cousin's seizure of Machecoul and Champtocé, but he had not immediately available enough troops to turn out the intruders, and he was forced to await the arrival of some fresh men, lent to him by Duke Jean V, who was naturally anxious about the fulfillment of his bargain over Champtocé. One would naturally have expected that, since the incriminating remains at Machecoul had now been destroyed, Gilles would march first on the other captured stronghold; but it is probable that he left Champtocé to the last, lest Duke Jean should withdraw his men, once his own interests were out of danger. Machecoul surrendered after a brief siege; and Gilles turned back to Champtocé, which also fell into his hands without much delay.

As soon as the place was his, the Marshal set about the destruction of any evidence against him. Choosing six of his most reliable retainers, including the two who had served at Machecoul, he swore them to secrecy in the most solemn manner, and then led them to their grisly task in the depths beneath a lonely tower of the castle. Two of them were lowered by ropes into the subterranean abyss, where they found themselves amid bones and decayed corpses. Meanwhile, Sillé kept watch outside, lest any of the troops should discover what was going on. The horrid remains were hauled up in a sack to ground level and packed into three huge chests. At least another 40 children, mostly boys, had come to their ends in that dreadful tower.

With this lamentable cargo, Gilles set out by boat down the Loire; the chests were put ashore at Nantes, and carted to Machecoul where a huge fire was set blazing in the Marshal's room. Some days were spent in destroying the relics of the miserable children; and during that time Giles wept copiously, beat his breast, and cried aloud to God for mercy. When all was over, and the last traces of his crimes destroyed, he ordered High Mass to be sung for the repose of his victim's souls.

Even this lesson failed to turn him from his evil course, and one must recognise that Manichaeism was rooted in the very fibre of his being. In his theology, God and Satan were co-equal and co-eternal; both were equally worthy of propitiation. And so the story goes on: children vanish while tending their sheep, others disappear from the street corner or while playing on the fringe of a wood; others are taken while returning from school; in La Suze not a male child is left; in Tiffauges not a child of either sex remains.

It is impossible to put a definite figure to the crimes of the Marshal. In Article xxvii of the indictment brought against Gilles by the Ecclesiastical Court, 140 or more cases were given. The Marshal himself is quoted* as putting the number of the murders at 120 per annum over seven years; but Mr. Vizetelly† regards this statement with not unjustifiable scepticism and thinks it "more likely that Gilles' victims were from 200 to 300." What we do know is that after his experiences at Champtocé and Machecoul, the Marshal took more pains to destroy the relics of his victims, once he had killed them, evidently fearing the discovery of the corpses if he continued his old careless methods; and thus the real total will never be known. In this place it is neither necessary nor desirable to dwell on the combination of cruelty, bloodshed, and unnatural lusts which marked Gilles' crimes.‡

We may now turn to the closing stages of the Marshal's career. In 1438, he despatched his chaplain, Blanchet, to Italy to enlist some expert who would help him in his spagyric and necromantic affairs. In Florence, Blanchet made the acquaintance of a young adept, Francesco Prelati, who seems to have combined a taste for high living with expertness in alchemy and the Black Arts; and in 1439 this last in the long array of Gilles' wizards made his appearance at Tiffauges.

Gilles received him with enthusiasm. At that date, Prelati was only twenty-three; but he was learned, witty, cultivated, adroit in

---

* *Curiosités de l'Histoire de France*, IIe Série: Procès Célèbres, par P. L. Jacob, Bibliophile (Paul Lacroix), p. 95 (1858).
† Vizetelly, *Bluebeard* p. 271 (1902).
‡ Those who desire to sup on horrors will find a light refection provided in Vizetelly's *Bluebeard* and Huysmans' *Là Bas*. A more complete menu is given in the Abbé Bossard's *Gilles de Rais, Maréchal de France, dit Barbs-Bleue* (1886), and in the reports of the Ecclesiastical and Civil Proceedings against Gilles.

flattery and skilful in the arts of deception; the very man for the Marshal. He professed to have himself seen a specimen of the Philosopher's Stone, so that there could be no doubt as to its existence. All that remained was to discover the method of preparing it. Again Gilles' hopes rose high. The alembics, pelicans, and athanors were brought out once more; the furnaces were rekindled; and the spagyrists plunged anew into their feverish quest, sometimes in the castle itself, sometimes in a neighbouring house. Nor did they depend on the methods of science alone. Blanchet, who was, after all, a priest, was shocked on discovering that Prelati and his master were seeking the aid of Satan, tracing magic circles, burning mysterious tapers, indulging in invocations and prostrations before the Prince of Darkness.

Once, while Gilles and Prelati were being surreptitiously watched by Blanchet in the act of supplicating Satan, a blast of icy wind swept down, which the priest regarded as a manifestation of the wrath of God. On another occasion, the two demonomists were engaged in an attempt to raise the Devil whilst Blanchet and Griart, the chamberlain, waited in an adjoining room; and on the roof over their heads they heard the mighty tramp of some four-footed monster hurrying to the summons. At another time Prelati, when braving the Fiend alone, was found by his friends to have been severely beaten and had to keep to his bed for more than a week. At one stage, their efforts seemed crowned with success. The Devil appeared before Prelati in the form of a handsome young man; and when the necromancer begged for wealth on behalf of the Marshal, the room was suddenly filled with ingots of gold. The magician hastened to bring Gilles, but when they reached the door of the room, Prelati, pushing it ajar, perceived a huge green serpent, "in girth as big as a dog." "Do not enter!" cried the magician, "There is a great serpent!" whereupon the Marshal took to his heels. When Giles plucked up his courage and went back to the room, there was no treasure. The Evil One had removed it, leaving only a few strips of gold foil behind. Little wonder that Giles was encouraged by such marvelous results. Prelati, also, must have been delighted to have fallen in with so simple a patron.

But Prelati's activities were not always so innocent. He joined the Marshal in his murderous practices and lent his encouragement in this field. And so the horrible kidnappings continued, interspersed with the Manichee's fits of remorse and repentance. In his days of contrition, he wandered over the countryside, gesticulating, crying incoherently, and terrifying the passers-by with his bedlamite behaviour. But then he would return in the evening to his castle, where he would carouse with Prelati and his intimates, drinking heavily in strong wine and devouring banquets of highly-spiced viands until his excited senses drove him to his horrible and murderous delights yet once again.

But by now even some of his retainers had fallen away. The major-domo, de Bricqueville, had gone. Blanchet, the chaplain, quitted Tiffauges and took refuge at an inn at some distance. Gilles tried to lure him back, telling him that Prelati's alchemical experiments were "going wonderfully well." But a chance meeting at the inn with the Castellan of La Roche-sur-Yon, Messire Jean Mercier, confirmed Blanchet's distrust of the Marshal. According to the Castellan the main topic of conversation over the whole of southern Brittany and Poitou was the conduct credited to Gilles. Among the country folk a rumour was afloat that the Marshal was slaughtering children to obtain their blood, and with this blood he was writing a magical book. As soon as this book was finished, ran the tale, not even the most powerful fortress would be proof against the Marshal, who would be invincible thereafter. This news made clear to Blanchet how far-flung was the feeling against Gilles, and he prudently decided to have no more to do with his late employer.

His late employer, however, had other views. He despatched his goldsmith to bring Blanchet back to Tiffauges. The chaplain refused, and repeated to Jean Petit the tales he had heard; and these the goldsmith communicated to Gilles when he returned to Tiffauges. The fury of the Marshal may be imagined, when he learned that his dearest hopes were the tattle of the countryside. He vented his rage on Petit, the goldsmith, sending him to imprisonment in the fortress of St. Etienne-de-Mer-Morte; and he despatched de Sillé and three companions to seize Blanchet, who evidently could not keep his tongue still. As his captors led him along the road, the chaplain realised that they were taking him to St. Etienne-de-Mer-Morte, to imprisonment, and perhaps to death. His entreaties prevailed on them to take him instead to Machecoul, which was only seven miles distant from St. Etienne-de-Mer-Morte; but he was warned by the valet Corillaut that any more blabbing would mean the end of him.

Now the ground was cracking under the feet of Gilles. The countryside was aflame with rumours of his butcheries and devil-worship. His own family harboured the gravest suspicions of him; for when René de La Suze seized Champtocé, although he did not discover the great charnel house, he found the dead bodies of two young boys concealed within the precincts; and probably only considerations of family honour kept him silent. Yet it was in these circumstances that the Marshal deliberately challenged the power of his overlord, Duke Jean V, and the relentless enmity of the Church. Madness could no further go.

Strangely enough, the details of the proceeding which led to the downfall of the Marshal are extremely obscure. St. Etienne-de-Mer-Morte was the last property on which he tried to raise money; and, as a result of this transaction, he cherished a grievance against a certain Jean Le Ferron, who seems to have been an intermediary for Duke Jean V, the real purchaser. Whatever be the grounds for the Marshal's grudge, he determined to pay it off; and on the morning of Whit-Monday, 1440, he arrived at St. Etienne-de-Mer-Morte accompanied by a small force of armed men. Jean Le Ferron was at Mass along with the rest of the townsfolk. As the ceremony drew towards its end, Gilles, with some armed satellites at his back, burst into the church, seized Jean Le Ferron and dragged him off to captivity at Tiffauges, along with some of his friends. This move was not without its cunning, since Tiffauges was outside the jurisdiction of the Duke of Brittany, and was held in fief from the King of France. But this minor cunning counts for nothing in comparison with the amazing blunder which Gilles had made. It was, indeed, a triple folly. At a single stroke he had violated the custom of Brittany which forbade a baron to raise troops without the express consent of the Duke; and he had committed a double sacrilege in profaning a church and in seizing Jean Le Ferron, who was a tonsured clerk and so under ecclesiastical protection.

The first result of Gilles' action was a summons from Duke Jean V, ordering him to release Jean Le Ferron and the other prisoners, and threatening a fine of 50 thousand crowns if he disobeyed. The Marshal treated this with contempt, retired to Machecoul, and awaited events. The exasperated Duke then seized St. Etienne-de-Mer-Morte and applied to his brother, Count Artus de Richemont, Constable of France, for assistance in dealing with Tiffauges, which was outside his jurisdiction. French troops were despatched to Tiffauges, and Gilles, perturbed by the course of events, surrendered his captives.

Richemont was a man of generous mind, loyal to old comrades, and ready to play the peacemaker. Gilles had served under him at the outset of his career, and the Constable had not forgotten. His skilful diplomacy smoothed over the quarrel between Duke and Marshal; no more was heard of the threatened fine; and Gilles set out with a full retinue to pay a formal visit to Duke Jean V at Josselin, some 20 miles north of Vannes. Prelati accompanied him to cast his enchantments and forewarn the Marshal of any danger.

It was a sinister cavalcade which wended its way across Brittany from Machecoul. His mind relieved by his reconciliation with the Duke, Gilles unreined his blood-lust with a complete disregard of risks. At Vannes, a ten-year-old boy was done to death at an inn and his body was sunk in a cesspool; at Josselin, under the Duke's very roof, several boys were slaughtered, according to Gilles' own confession; and at Bourgneuf-en-Rais, on the return journey, Bernard Le Camus, a boy of fifteen, was murdered—the last known victim of the Breton Minotaur. His last victim; for now Justice, lame and slow, but sure, was close on the heels of the Marshal. He returned to Machecoul, with its frightful memories. On the evening of September 13, 1440, an armed company appeared at the castle gates, the escort to Robin Guillaumet, process server to the Bishop of Nantes. Gilles had made his peace with the Duke; but now he was to face a relentless enemy: the Church.

At this juncture, Jean de Malestroit, Chancellor of Brittany and Bishop of Nantes, became a principal actor in the drama. He had old grudges against Gilles, and more serious grounds for umbrage against Richemont, who had once thrown him into prison as a traitor. He had great influence over Duke Jean V, whom he resembled in character, for both were avaricious, long in foresight and patience when personal interests were at stake. And now Malestroit saw his way to strike at Gilles and, through Gilles, at the hated Richemont. By his double sacrilege at St. Etienne-de-Mer-Morte, the Marshal had brought himself within the grasp of the Church, and Malestroit meant that full payment should be exacted. Even then, however, he had not yet realised the full weakness of his opponent.

Gilles' quarrel with Duke Jean V served to unleash the tongues of the populace, since they believed that their hated master was tottering towards a fall. Tales reached the ears of Malestroit which must have spurred him to action. Even while Gilles was paying his ceremonial visit to the Duke, the Bishop had thrown himself feverishly into an investigation of the Marshal's crimes. He quitted his palace in Nantes, scoured the district, examined witnesses, accumulated evidence which must have appalled him, enemy of Gilles though he was. Then, with the facts before him, he sent his process server to Machecoul.

Gilles seems to have been thrown off his balance. His habitual counselors, de Sillé and de Bricqueville, getting wind of coming trouble, had decamped already. Prelati would have followed them if he could. A cold fit came over the Marshal, and he surrendered to the little force at his gates. He was seized, along with some of his accomplices, taken to Nantes, and thrown into prison. Machecoul was searched; and in the house occupied by Prelati and Blanchet were found some ashes—believed to be human remains—and the blood-stained shirt of a little child.

The arrangements for the Marshal's trial were complicated, for both ecclesiastical and civil powers took their share, and in addition a representative of the Inquisition was adjoined. Jean de Malestroit presided; three Bishops assisted him; the Vice-Inquisitor sat beside him; there were representatives of the civil power; and a host of legal experts. The ecclesiastical proceedings occupied a month and eight days. Then, since *Ecclesia abhorret a sanguine*, the criminal was handed over to the civil power, who finished their part of the business in 48 hours.

There is no need to follow the trials in detail. When it sat, the ecclesiastical court met early in the morning, since it was the practice that judges and witnesses should fast until the day's evidence had been heard. On September 19 Gilles was charged with heresy only; and he admitted the competence of the court, declaring that he was prepared to prove his innocence. At this stage, he does not seem to have realised that his life was in danger. During the adjournments he spent much of his time at Mass, though the priest was forbidden to listen to his confession, absolve him, or permit him to communicate.

On October 8 a fresh indictment was put forward in which he was charged with murder and vice, in addition to heresy. Gilles, quite properly, pointed out that the court had no jurisdiction in civil matters such as murder, and refused haughtily to recognise its authority. On October 13 an indictment containing no less than 49 articles was read; and Gilles burst into fury, denouncing his judges and protesting against the procedure. The answer of the court was to pass sentence of excommunication upon him.

In those days, excommunication was a dreadful weapon; and it broke the Marshal's courage. When he reappeared before his judges on October 15 all his bravado had vanished; he meekly submitted to the jurisdiction of the court and prayed for pardon for his previous violence. Then his accomplices, Griart, Corillaut, the woman La Meffraye, Blanchet, and Prelati, were brought in to give their evidence; and Gilles was asked if he wished to cross-question them. He replied that he trusted to their truthfulness; and they were removed in order to have their evidence taken down in writing by the appointed clerks. But their mere appearance sufficed. No sooner had they left the court than Gilles broke down completely, weeping and beseeching on bended knee that the sentence of excommunication should be cancelled. Once again he had swung from one extreme to the other; and as his contrition was plain, the excommunication was withdrawn.

On October 20, the evidence of the more important accomplices was read out in court, and Gilles was asked if he had anything to object against it or the witnesses themselves. To this he answered "No," which was tantamount to confession. But this was not sufficient in the eyes of the Church. The ecclesiastical prosecutor—the Promoter, as he was termed—pointed out that a confession could not be accepted if it were "*dubia, vaga, generalis, illativa, jocosa,*" and he demanded that for certainty Gilles should be submitted to the "canonical question," i.e. to torture.

Gilles had already given way under excommunication: at the threat of torture he collapsed completely and offered a full confession in private, which he was prepared to avow later in a public session. He was therefore examined privately by the Bishop of Saint-Brieuc, representing the Church, and Pierre de l'Hospital, President of the Breton Parliament, who represented the civil power. At one stage, Prelati was brought in to testify; and when his accomplice was being removed, the Marshal made him an affecting farewell, bidding him "Rest assured that if you have good patience and trust in God, we shall see each other again amidst the great joy of Paradise." The Manichee, obviously, had swung over completely now into the camp of the Deity.

Meanwhile, the civil court, under Pierre de l'Hospital, had also been dealing with Gilles. When he first appeared before it, the Marshal was still unconscious that his life was in peril. He begged his judges to hasten their proceedings, as he was anxious to bestow rich gifts on the Church, give the greater part of his wealth to the poor, and, by dedicating himself to the service of God, ensure the salvation of his soul. The President dryly pointed out that while it was well that he should think of his soul, human justice must be satisfied also. Let him listen to the indictment. It included, among other civil matters, the charge that Gilles "did take or cause to be taken many little children, not only ten or twenty, but thirty, forty, fifty, sixty, one hundred, two hundred and more, and, indeed, so many that one can make no positive declaration of their number .

. ." and "did inhumanly murder and kill them, afterwards burning their bodies to convert them into ashes."

This was on October 11, when Giles was still in a mood of defiance, and he contested several of the terms of the indictment, denying especially all crimes against children. Asked if he would accept the evidence of his retainers Griart and Corillaut, he made the staggering assertion: "I received only honest folk in my house and service. Had I known any to be evil, I should have been the first to lay my hands on them." The civil court had other ideas about honesty and evil, for on October 23 it condemned these two worthy fellows to be hanged and burnt.

On the previous day, the Marshal had made his confession in open court. It was, apparently, an eloquent effort in which the recital of almost incredible horrors was mingled with obviously sincere expressions of repentance, and it ended with an appeal to the "relatives and friends of the children whom I did so cruelly put to death" to grant him the assistance of their prayers.

The court adjourned until October 25. On that day, Gilles was brought before the ecclesiastical court to hear its sentences, for there were two of them. The first was that delivered by the Bishop of Nantes and the Vice-Inquisitor jointly: it condemned him for apostasy, heresy, and devil-worship, and pronounced a fresh excommunication on Gilles. The second, by Malestroit alone, was on the grounds of vice, sacrilege, and violation of the privileges of the Church, and by it he was excommunicated yet again and order given that he should be punished for his salvation's sake. Gilles pleaded for the withdrawal of the excommunication. This was granted, and he was allowed to make his confession and was given absolution. The Church had finished with him. It handed him over to the civil power.

The same evening, he appeared before the civil court, dressed in black; and again he made public confession of his crimes, before an immense audience who had come to see the end of the "Exterminating Beast." After consultation with his assessors, Pierre de l'Hospital pronounced sentence. For his felony against the Duke, he was to pay a fine of 50 thousand crowns; for his other crimes he was to be hanged and burnt, at eleven o'clock on the following morning.

Giles had a request to make: that he and his two accomplices might all die at the same hour. "I am the cause of, and the principal in, their transgressions. . . . I can set them the example of dying well. For if it were otherwise, if my servants should not see me die . . . they might imagine that I should remain unpunished, I who am the cause of their crimes." This was granted. The Marshal asked one final favour: that the judge should petition the Bishop of Nantes to ordain that the execution should be preceded by a General Procession to pray God for him and his accomplices and fortify them in their hope of salvation. This also was granted.

Next morning the population of Nantes, after attending Mass, joined in the General Procession and the air was filled with prayers and chanting. One may be forgiven for thinking that the most suitable music would have been the penultimate movement in Berlioz' *Symphonie Fantastique*. But Gilles, to do him justice at the last, "died well." He addressed heartening words to his accomplices. The Manichee made his final swing in a confidence of salvation which leaves the ordinary mind astounded. "Love God," he adjured them, "and feel such regret for your offences that you may not fear the death of this world which is, indeed, but a little departure, without which one may not see God in His glory." Griart and Corillaut responded in very seemly terms. Then, as a last request, Gilles begged that he might be put to death before his accomplices, so that they might derive courage from his example. This, too, was granted.

And so, at the age of thirty-six, died Gilles de Rais, in the perfect confidence of salvation.*

---

* Michelet, in his *Histoire de France*, deals with the career of Gilles de Rais. J. K. Huysmans devotes a considerable portion of his novel *Là-Bas* to an account of the Marshal; but he is hardly to be trusted in all details. More reliable is the Abbé E. Bossard's *Gilles de Rais, Maréchal de France*,

Attempts have been made to connect Gilles de Rais with the Bluebeard of Perrault's fairy tale; but obviously Perrault's story has no direct connection with the career of the Marshal, who married only once and whose wife survived him. After the publication of Perrault's fable, popular fancy in Brittany appears to have associated the names of Gilles de Rais and Bluebeard. A careful analysis of the problem will be found in Vizetelly's *Bluebeard*.

In conclusion, one more quotation may be given; it is taken from the public confession of Gilles. "Fathers and mothers who hear me, and you friends and relations of young folk whom you love, I beg you keep watch over them. Mould them by teaching them good principles, good examples, and healthy doctrines. Nourish their hearts with these, and above all things fear not to correct their faults, for, were they reared as I was reared, free to do as I pleased, they, perchance, might slip likewise into the same pit." Gilles was, perhaps, no expert on education, but at least he spoke from bitter experience. Some modern educationists, please note.

(cont.) *dit Barbe-Bleue*. There are short accounts of Gilles in Brevannes's *L'Orgie Satanique à travers les Siécles* and in M. A. Murray's *Witch-Cult in Western Europe*. A brief, rather one-sided notice of him is to be found in the *Encyclopaedia Britannica*. The best life of Gilles in English is that given by E. A. Vizetelly in his *Bluebeard*, which contains a full story—so far as that is possible in our language—of both Gilles de Rais and his earlier exemplar, Comorre. Mr. Vizetelly's work is based on the original documents, and he has corrected sundry errors of earlier writers on the subject. In the foregoing sketch I have depended on Mr. Vizetelly's account for many of the facts about the Marshal, and I wish to make acknowledgement of my indebtedness in this respect.

X
A PRESENT FOR A GOOD HISTORIAN

WERE YOU PROPERLY BROUGHT UP in your nursery? Did you read *Winnie-the-Pooh*, or *The Wind in the Willows*, or *The Just-So Stories*, or *Alice in Wonderland*, or *The Water-Babies*, or *Gulliver's Travels*? If you read none of these, then you were a sorely-neglected child and should bear a just resentment against those who had charge of your early education.

I am afraid that nowadays *The Water-Babies* resembles the King, whom we all know by name but with whom only a few of us have a personal acquaintance. In my young days, we rubbed shoulders with Kingsley's book, and in it I encountered my first criterion for differentiating Man from Beast. "Nothing," I learned, "Nothing is to be depended on but the great hippopotamus test. If you have a hippopotamus major in your brain, you are no ape, though you had four hands, no feet, and were more apish than the apes of all aperies. . . . Always remember that the one true, certain, final and all-important difference between you and an ape is, that you have a hippopotamus major in your brain, and it has none."

I was much perturbed. Was there room, in a small skull like mine, for such a formidable thing as even an ordinary hippopotamus, not to speak of a hippopotamus major, which must be larger still? My father reassured me. If I had no hippopotamus major in my brain, I had a hippocampus minor, which was just as good a proof of my humanity.

It was years later before I learned the details of the great controversy between Huxley and Owen over the hippocampus minor, which is "a small eminence, shaped more or less like the sea-creature of that name, in the backward prolongation of the central hollow of the brain technically termed the posterior cornu of the lateral ventricle." Sad to relate, if I were shown a brain, even to-day, I should be unable to tell whether it contained a hippopotamus major, or a hippocampus minor, or both, or neither: for I am no anatomist. So, for my own satisfaction, I have had to seek some other means of distinguishing between Man and his kindred animals.

Now baboons possess a social system and have a certain aptitude for launching projectiles; the elephant displays wisdom; some beast-packs follow a leader; the cat on the tiles can make himself comprehensible to his lady love by means of his voice, and what more could Propertius hope to do for Cynthia? Characteristics such as these are no acid tests to separate Man from Beast.

There is, however, one diagnostic which will serve. Through countless generations, the ancestor of modern man competed with his fellow-animal as one beast against another. Then, in history, came a cleavage, and thereafter a chasm divided humanity from the rest; for at that stage man learned to control fire, a feat which no beast has ever achieved.

It was not the *discovery* of fire which set the human animal apart from its neighbours. Long ere that segregation took place, fire must have been known to many creatures through eruptions of volcanoes, lightning-kindled conflagrations, and self-generated forest fires. But, whilst the brutes feared the flame and shunned it in horror, some man must have been bold enough and intelligent enough to experiment with the dreaded phenomenon, to learn how it could be fed, transferred from place to place at need, or extinguished at will. Once that control was achieved, man had his foot on the long road to supremacy upon the surface of the planet. He had gained what no other animal possessed: a protection against the terror by night, the peril of the sabre-toothed tiger, the giant hyena, the pitiless wolf, ancestral memories of which, perchance, still make nightmares of our dreams.

Thus, at the birth of *homo sapiens*—as distinct from the animals which preceded him—we come upon a chemical reaction, the control of which changed the whole current of history on this planet. Before the human animal learned the management of fire, he was a beast among beasts, filling his maw with fragments of raw and bleeding flesh rent from the carcase of some victim; unable to store his food safely from the ravages of insects or fungi; and dependent for his defence upon the weapons which he could contrive from wood and stone. Look, by contrast, on his position after he had mastered fire.

Primitive cookery can have been no dainty business; yet in the gruesome singeings and scorchings round the camp-fires of the Stone Age lay the germ of those refinements which delight the palates of Brillat-Savarin's successors. Our methods have improved with the passing of the generations, but the principle is still the same: to induce a chemical reaction in the raw material by the aid of heat.

An analogous transmutation in the structure of clay gave rise to the art of the potter; and one of the results was the designing of those huge jars which Schliemann found in buried Troy, wherein grain could be stored in safety, protected against the ravages of insects and of damp. So came the possibility of building up a reserve against a time of need, and a stability in the social system which must have been absent through the previous ages when a bad season might reduce a tribe to starvation.

Even greater was the advance in the fields of tools and armaments. By skilful charring and hardening, a stake could be converted into a formidable weapon. Then came the beginnings of metallurgy, the age of copper, the age of bronze, and, thereafter, with increased skill, the age of iron.

The opening of the brief age of copper, and its longer successor the bronze age, confronts us with problems upon which we can only speculate. How did early man hit upon the extraction of copper from its ore? One is almost forced to the conclusion that by accident some primitive tribe built into its hearth, among other stones, some chance-gathered copper ore; and that, after the fire had burned for days, traces of the metal were found among the ashes.

Anyone who imagines that primitive man was a creature of low intelligence may be recommended to ponder on this matter. The properties of this heaven-sent gift marked it out from anything with which its discoveries were previously acquainted. It may have been recognised as akin to gold; for gold is found native in the metallic state, and some tribes must have been cognisant of the malleability and sheen which sharply differentiate it from ordinary stones and rocks. In any case, here was something of immense practical value: malleable, yet far harder than gold. But to trace the origin of it to an ore of entirely different properties and appearance, to identify one particular kind of stone amongst the others which formed the hearth, and to connect ore and copper with the action of fuel and fire: these are operations demanding intelligence of a high order.

Merely because we have inherited the accumulated knowledge of a thousand generations and are thus able to produce wireless and aeroplanes, it would be ludicrous for us to despise these humble discoverers who blazed the trail for us in the days before recorded history began. They must have had brains as good as our own, though naturally they lacked the knowledge granted to their successors. There were mighty men before Agamemnon; and for my own part I prefer the *Odyssey* to *Paradise Lost*, despite the fact that Homer lived in a misty past so remote that Milton, by comparison, seems almost our contemporary.

Now just as fire acted as a reagent to separate man from beast, so did this new-found craft of metallurgy serve to favour the bronze-armed warrior over the slower-witted tribes who clung to wood and stone, and the man of the Iron Age against the laggard who adhered to bronze when its day had passed. Thus, the chemical reactions of metallurgy, all uncomprehended though they were by those who used them, served as a sifting-machine which threw aside the backward portion of mankind and gave supremacy to the more cunning.

Take another early discovery, the process of preparing wines and ales from fruits and grain. Fermentation is merely a chemical reaction in the course of which alcohol is produced from the carbohydrates of plants. And in alcohol, to some extent, we find another of these sifters of the human community. The weaker brethren succumb to it, lapse into drunkenness, and become worthless or even deleterious; those with stronger wills or more resistant physiques take no harm and persist.

But if one wishes to study the social influence of the fermentation reaction on a large scale, one need but glance at the results of the Prohibition experiment in the United States. Before it was discontinued, it produced in full flower that anti-social type, the gangster; and it is reasonable to suppose that the Fascists and the Nazis learned some of their technique from the Transatlantic bootlegger and hijacker. Clearly the social results of some simple chemical reaction may be nation-wide and in their repercussions may lead not only to the breaking-up of ill-conceived laws, but to red ruin as well.

Let us now turn to the problem of dwellings. In the early ages, we find mankind huddled in caves which formed the only secure shelter for a race at that stage of craftsmanship. Then comes a migration to villages built on piles above the surface of lakes, where the still impuissant humanity could feel safe from the onslaughts of the formidable carnivores. Later, like the ancestral amphibian, man moved out on to the land, when his better armament secured him a safe footing. But there, in the early stages, his shelters must have been miserable: a wigwam, or some wretched wattle-and-daub hovel, or at the best a hut built of rough stones secured together with plastered mud.

In such conditions, nothing larger than a hamlet could exist in the open country, removed from rivers and lakes; for the problem of water supply alone would raise insurmountable difficulties against the formation of a larger unit. Also, with a growth in the perimeter of an area, the difficulties of its defence may increase unless the site be well selected.

Yet the population grew, and social machinery became more complex when hunting tribes settled down to agricultural and other stable and specialised pursuits. The earlier forms of habitation ceased to meet the changed conditions. Something at once more compact and more permanent was needed.

Once again, an uncomprehended chemical reaction proved to be the saving factor. The discovery is so ancient that neither the name of the inventor nor even the manner in which he made it, has come down to us. Doubtless some limestone found its way into the wall which hemmed in the tribal fire, and lime was formed accidentally in the heat of the flames. Then, when the embers had grown cold, a shower of rain may have slaked the quicklime; the rising steam may have caught the attention of some tinker and led to further investigations. Somehow or other, mortar was discovered. At a later stage, the economy effected by mixing it with rubble was recognised, and thus concrete was invented. Two new building materials were added to the growing resources of humanity.

As a sequel to these discoveries, a fresh type of dwelling came into being. Instead of frail, squat shanties, inadequate in the face of wind and rain, there now rose, storey above storey, the edifices of the newer age, spacious, well-roofed, and solidly-constructed to withstand the batterings of storms or of human foes. The art of architecture entered into a new phase. Palaces arose. Fortresses loomed over the countryside, with the social results which can be studied in the history of the Feudal Age.

Another effect of the employment of the new building materials was the rise of close-built cities, housing large populations in a restricted compass and suffering all the social adjustments demanded by such conditions of life. If a town or city outgrew its local water supply, aqueducts could be constructed to carry the fluid from far-distant lakes and thus give opportunity for a further safe expansion of the population. Or if the acreage suitable for building was restricted by the environment, the city buildings could be made to extend upward instead of superficially. Thus, to that unknown discoverer of the lime-slaking reaction, New York owes the possibilities of its upward growth towards the sky.

The foregoing paragraphs indicate some of the knowledge which spread among mankind during early periods—but in addition there must have been a very wide range of information of other sorts: the edibility or toxicity of certain plants, the seasons most suitable for sowing and harvesting, the methods of constructing ponds for catching and storing water, the best mechanisms for trapping animals, and so forth.

Up to a certain stage, the total knowledge of a tribe was so limited that it could be passed in its entirety from father to son, either orally or by practical illustration; but eventually the range of information grew until these methods broke down. The older generation failed to transmit, or the younger one to assimilate, all that had been learned; and so stagnation or even retrogression might ensue. If all the steadily increasing mass of fresh information was to be added to that already stored up in reserve, some means of record was required which was more reliable and more permanent than human memory.

We know some of the methods devised to this end: the knotting of cords, the notching of wood, the winding of strings about twigs, and inscriptions cut in stone. In Mesopotamia, the cuneiform writing produced by impressing soft clay tablets with a pen in the form of a triangular prism gave birth to huge libraries; but even a *vade mecum* constructed on this system would require a wheelbarrow or a truck for its transport. Something more readily portable was demanded.

The germ of our modern solution of the problem arose in ancient Egypt where, by slitting and flattening the casing of the papyrus reed and joining the slips together by pressure, something akin to our paper was devised. But, in addition to a receiving surface, a pigment was required in the form of ink; and here chemistry again appears on the scene. The earliest inks consisted of lampblack compounded with gum, which made the black particles adhere to the surface on which the writing was inscribed. Then came the discovery that when an iron salt was mixed with extract of gall-nuts, it gradually darkened on exposure to air and gave permanent markings. But inks of this description were very pale during the actual writing; and when aniline dyes were discovered during last century, it was found convenient to add a blue dye to the pale iron-gall solution and so produce a mixture which was deep enough in colour to make the ink clearly visible at the moment of writing, whilst permanency was assured by the gradual oxidation of the iron ink by the atmosphere.

Concurrently, the receiving surface underwent its own course of development. As the art of writing spread across the world, some material more widely distributed than papyrus was demanded. Sheepskin was employed; but before the raw hide was fit to inscribe, it had to be subjected to chemical treatment. Then came the conversion of linen rags into paper. Finally, chemistry showed the way to convert wood-pulp into a form of paper which has come more and more widely into use in recent years.

If one remembers that the paper in a single issue of a well-known newspaper weighs 100 tons, it gives some idea of the amount of deforestation required to keep the daily press of this country running during a paper-shortage; and in normal times the tree-cutting would be on a far vaster scale, and might even have an ultimate effect on the climate of the regions from which the wood is drawn. Further, the syndicated press provides the opportunity of disseminating the ideas of the staff into the minds of the public on a gigantic scale; in some cases, where the readers take their ideas entirely from the press, the social influences thus exerted may be potent indeed. And all this is made possible by a simple chemical reaction which makes the supply of cheap paper available in huge quantities.

Let us turn to another series of phenomena which have not been without their influence on human life. It is not known who first observed that when sulphur and charcoal were mingled with nitre, a mixture resulted which was capable of deflagration; but it seems to be a discovery of respectable antiquity, since the fact was known in China long before the Christian era. As a weapon, gunpowder seems to have come to the knowledge of Europe via the Saracens; but it was not until the fourteenth century that it was employed on any scale in European wars. Later, it was found to have uses in peace as a blasting agent. With the advance of scientific investigation, the original gunpowder has been displaced by a horde of more powerful and more easily-manipulated nitro-compounds.

Even in its early days, however, gunpowder exerted a certain social effect. Previous to its introduction into war, the mailed knight stood in a class by himself, far less vulnerable than the man-at-arms in his leather jerkin; but with the introduction of the arquebus, the steel-clad warrior was no safer in battle than his humbler acolyte. Armour, and all that armour implied in visible caste-distinction, receded into the background. The feudal castle, too, which had been almost invincible against the attacks of bowmen, found itself insecure when the bombard came into action against it; and the decline of the formidable citadel worked a vast change in the social life of the countryside. The Black Death played the major part in the fall of the Feudal System, but it would be unjust to deny gunpowder its share in the credit for that overthrow.

In peace, gunpowder had its uses in economising the human labour of quarrying, mining, and, tunneling, and in the blasting of dangerous rocks from channels. Its successors, the nitro-explosives have proved even more efficient in these fields.

But to see the full social influence of explosives nowadays one need but look at the devastation of the European cities and reflect upon the problems which this produced in the days of the evacuations and upon those others which are still to come when the time is reached for the rebuilding of the damaged cities and towns. It is a far cry indeed from this to the discovery of gunpowder, but the one has grown from the other as the oak-tree springs from the acorn; and it is a poor perspective which looks solely at ultimate consequences whilst ignoring the germ from which they sprang.

The Black-out has given people of the present day some inkling of what life was like in the streets of the eighteenth century by night; but our social standards have risen since the days of Queen Anne, and except for sporadic incidents we had little to fear in our highways during the period of darkening.

Very different were the conditions in the London streets of the eighteenth century, when watchmen perambulated at dusk, crying the warning: "Hang out your lights!" In those times, the only available means of street-illumination were the dim and flickering oil-lanthorns pendent from the householders' lintels. Highways were lost in gloom; a penetration up lanes and alleys was an adventure full of hazard. In such conditions, efficient policing was impossible; cut-purses lurked in safety under the cover of the darkness, waiting for their victims; and to these professional scoundrels were added the wandering bands of aristocratic Mohocks seeking their sport in molesting and beating the unprotected passer-by. Little wonder if the ordinary citizen preferred to remain after nightfall within the safety of his own house and to go abroad in darkness only in groups, so that numbers ensured a certain protection. With the passing of twilight, the pulse of the city's social life beat more feebly. Compare with this the modern peace-time nocturnal throngs in well-illuminated streets and byways, the inter-visiting of friends after dark, the tide of amusement-seeking individuals crowding the thoroughfares at theatre-time, the packed restaurants and night-clubs, and the general change in our social habits induced by this increase in safety.

The root of this momentous transformation in social customs lies in a single chemical invention: the generation of gas by the distillation of coal.

It is a commonplace that when a material is produced cheaply in order to fill one particular need, other uses will soon be found for it, and coal-gas was no exception to the rule. With a plentiful supply of this convenient fuel, attempts were made to employ it in fields other than lighting; and the gas-engine was invented, one of the most momentous devices ever contrived.

But the gas-engine, in its original form, was tethered to its source of supply, and before it could exert its full effect upon civilisation it had to undergo the modification which converted it into a locomotive machine. The principle remained the same, but instead of coal-gas the new fuel was petroleum, which also furnishes an explosive mixture when mingled with air.

In this manner, the original gas-engine became transformed into the modern motor-vehicle, carrying its own compact fuel in its tank and capable of long uninterrupted journeys on the roads of the countryside. Yet, without the aid of another chemical discovery, we should not enjoy the speed and comfort of the present-day car.

Rubber, in its natural state, is too soft a material to stand hard wear; but by subjecting it to the action of sulphur a new substance, vulcanised rubber, is obtained, which is far tougher and more resistant. Thus, the chemical discovery of the vulcanisation process opened the way to the use of rubber on the tyres of vehicles, and a primitive shock-absorber was to hand, which increased both the speed and the comfort of passengers. Later, the solid rubber tyre gave place to the cushion tyre and it, in turn, was displaced by the invention of the pneumatic tyre. But without the basic discovery of the vulcanisation process the modern rubber tyre would never have been possible.

Thus, upon two chemical discoveries (the explosiveness of a mixture of petrol and air and the vulcanisation of rubber) hangs the whole of that modern system of transit which has already altered our civilisation so profoundly, and which may yet change it still further in the future. Lorry-traffic now competes with the railways as, in an earlier day, the railways competed with the canals. Goods can be transported in quantity into the most remote spots where they may be required for use. Motor-buses are changing the distribution of our population and bringing a mobility which was not even dreamed of fifty years ago. On the sea, the motor-vessel is beginning to run side by side with the steam-propelled ship. And in the future the pace will grow yet faster, as we can infer from the advance of the aeroplane, which has already established itself as one of the most potent weapons in total war.

Two world wars have brought home to the average citizen the importance of fats in his diet, so that it is unnecessary to stress the significance of the margarine industry nowadays. Yet it is only about seventy years since the first attempt was made to produce a palatable butter substitute. When acetylene gas is mixed with hydrogen, no combination between the two gases occurs; but if certain substances, such as finely-divided nickel, are present, they act as catalysts and the hydrogen unites with the acetylene at quite moderate temperatures. In a similar manner, hydrogen can be forced to combine with numerous natural oils; and by this means materials are formed which make passable substitutes for butter.

Outside the Polar regions, no one would care for a diet of whale-oil; but whale-oil, by hydrogenation, can be converted into something akin to ordinary butter. Coconut oil, cottonseed oil, and oil of sesame lose their individual flavours by hydrogenation and are thus converted into edible fats. It is not rash to predict that in future the world's need for vegetable oils may strengthen the position of those powers which control the Tropics and which can thus regulate the export of tropical oils to their less fortunate neighbours. Wars in the past have originated in much less important matters.

Another aspect of the food problem deserves more than a glance. Members of the animal kingdom subsist either on the flesh of other animals or upon the plants. They are unable to assimilate, direct from the soil, such simple chemicals as ammonia, carbonates, nitrates, phosphates, and water, but must have these things built up for them by the plants into more complex substances like proteins and carbohydrates. If all flora perished from our planet to-morrow, mankind and its kindred would be extinct after the briefest span, since no method has yet been devised for synthesising the complex compounds which constitute the only diet upon which we can nourish ourselves. Thus, in a sense, every animal is a mere parasite on the plants which act as hosts to supply it with nutriment.

Suppose a wheat crop is grown on a field. Nitrogen compounds are withdrawn from the soil to build up the proteins of the plant; and these proteins are eventually absorbed by the animal which eats the wheat. Thus, nitrogenous compounds will be steadily abstracted by wheat-crop after wheat-crop until the soil grows so poor that it cannot sustain any more wheat-growing.

To revivify the ground, it is necessary to enrich it with fresh nitrogenous compounds; and in the past the nitre beds of Chili have been drawn upon to the extent of hundreds of thousands of tons per annum. But this, obviously, merely throws the problem one stage further back, since sooner or later the Chili deposits are bound to be exhausted. Consequently it was essential, before this catastrophe actually occurred, to find a fresh source of combined nitrogen.

Chemistry again came to humanity's rescue. By using appropriate reactions, it is possible to force the nitrogen and oxygen of the atmosphere to unite with each other; and further treatment yields nitric acid and nitrates which plants can assimilate. Also, by catalytic processes, ammonia can be synthesised from hydrogen and atmospheric nitrogen. Thus, thanks to these chemical processes, the practically inexhaustible supply of nitrogen in the atmosphere can be tapped and converted into the food-stuffs of plants; and so the exhaustion of the Chili nitre beds is no longer a danger threatening complete disaster.

If dividends were a test, then two of the most important activities of humanity would be the purveying of food and the supply of amusement to the populace. The most cursory examination will reveal the growth during the last century in the popularity of certain recreations such as theatres, music-halls, horse-races, dog-races, football matches, the radio, and the cinema; and it seems interesting to note that all of these belong to what may be called the collective and standardised type of amusement—things which may be enjoyed simultaneously by thousands of individuals, and which conform to certain stereotyped patterns. It would be interesting to speculate upon the ultimate effects of this systematisation of public recreation, but that would overleap the bounds of the present subject, and it will be sufficient here to indicate how chemistry has contributed towards one of the most widely-disseminated forms of this kind of mass-amusement—the cinema.

Photography originated as a method of recording silhouettes and the structure of leaves by the action of light upon paper soaked in silver nitrate. Then came the Daguerreotype, with a metal plate as a basis. The Talbotype process followed, in which the first "negatives" and "positives" played their part. Then came the collodion "wet-plate" method, which demanded the use of a portable dark room. These processes were all much too laborious to attract any but the most enthusiastic amateurs; but in 1871 a gelatine dry plate with a glass backing was invented; and thereafter photography as a pastime spread into wider circles.

Meanwhile chemistry had added a fresh material to the resources of civilisation by the invention of celluloid, which was obtained by mixing gun-cotton with camphor. This substance was employed as a backing for the sensitive film in photography instead of glass, as it had the advantages of lightness and freedom from fracture by accidents. In 1889, Eastman put the first roll-film on the market and followed this, two years later, by inventing the "daylight loading" principle, two improvements which brought photography within the reach of even the least skilful and floated it into the popularity which it reached among the general public.

So much for the chemical side for the present. In the optical field, the toy known as the Zoetrope, or Wheel of Life, was devised in 1830. By means of it, a series of pictures could be presented to the eye at fixed intervals in such a way as to produce the illusion of a moving object: a ballerina dancing or a horse cantering. Edison, in 1890, blending together the two inventions of the Zoetrope and the roll-film, thus contrived the first cinematograph. Finally, "sound-tracks" were recorded on the margins of the pictures, and the modern "talkie" came to birth.

In the earlier days of cinematographic exhibitions, however, there were not infrequent and serious disasters due to the violent inflammability of the celluloid films; for a substance prepared from gun-cotton and camphor was very liable to catch fire during its passage through the lantern, and, once alight, it was one of the most difficult to extinguish. Again chemistry was set to work, and the modern non-inflammable film was the result.

From the foregoing, it will be seen how large a part the chemist has played in the perfecting of the cinematograph, first in the development of the photographic side and then in the invention and improvement of the celluloid backing for the sensitive film. The results of this long and laborious series of researches confront us everywhere to-day in the picture houses which have been scattered all over the globe. Here we have the best example of that curious standardisation of amusement which is one of the marks of present-day civilisation. But, in addition to its service to public recreation, the cinema furnishes one of the most subtle means of propaganda yet invented; and its effect in this field is a thing which no student of our modern civilisation can afford to neglect. All successful propaganda rests on the principle of telling the truth— but not the whole truth; and since we remember sights better than arguments, a carefully-designed cinema film may leave a stronger impression than any letterpress. Add to this the ubiquity of the cinema, and its influence can be appreciated. But this does not comprise all the results derived from the cinematographic camera. The sudden shortage of ordinary films which made itself felt in the early days of the war is the surest indication of the extent to which aerial photography has thrust itself forward into the military sphere nowadays as a means of unbiased observation and recording of enemy activities or vital topographic details. Thus, even the most cursory consideration suffices to reveal the immense influence which photography has exercised during its brief existence.

There is another field in which chemistry has played a part in modifying, the social outlook and mingling the classes so far as superficial matters go. In the second world war there has been a revival of something which dates back to the days when Greece and Rome were in their prime: legislation intended to regulate the private expenditure of citizens. The Lacedæmonian was forbidden by sumptuary laws to own a house or furniture built with anything more elaborate than axe and saw. Under Solon, there were enactments against extravagance in female attire; and in Rome, under the Oppian laws, no woman might possess more than half an ounce of gold for ornament or wear a dress of different colours. The Imperial purple was reserved for the ruling group and might not be worn by lower grades in society; and the trace of this privilege persists even to-day in our phrase "born in purple."

In England, under Edward III, a similar system prevailed. Wives and daughters of servants might not wear veils costing more than twelve pence; the wives of yeomen or craftsmen were forbidden to wear silk veils; fur was permitted to the ladies of knights with a rental of 200 marks per annum. The wife of a labourer was forbidden to wear clothes costing more than a fixed sum, and she could not wear a silver-garnished girdle. The wife or daughter of an esquire or gentleman might not wear velvet, satin, or ermine. A knight-bachelor's wife was forbidden to wear velvet; and the wife of a knight could not venture to don a cloth of gold or sable fur, which were reserved for classes still higher in the social scale.

Now, although the sumptuary laws died out in practice, the social differentiations between the classes in the matter of dress persisted, right up to the early years of the twentieth century. During the Industrial Revolution, a financial chasm opened between the low-waged labourer on the one hand and the employing class on the other; and as a result, the former contented himself with homespun garments and left silks and satins to the members of the moneyed group. Thus, up to the beginning of the twentieth century it was possible, even at a distance in the street, to classify the social standing of a passer-by without too much risk of error, merely by examining the clothes. In the case of women, this was especially easy.

But here again chemistry has stepped in, this time as an equalising agent. As far back as 1886, a commercial process was invented by which cotton was nitrated and the resulting gun-cotton was dissolved in a mixture of ether and alcohol to form collodion. By forcing this collodion through fine jets into water, a fibre was produced which resembled silk. This is denitrated by treatment with an alkaline hydrosulphide.

Another process also uses cotton as its basis. The cotton is dissolved in a liquid prepared by the action of ammonia upon oxide of copper ("cuprammonium solution") and the liquid is driven through spinnerets into acids or alkalis, whereby the cellulose is precipitated in the form of fibres. Two other processes have wood-pulp as their raw material, and are therefore more economical. In the one, wood-pulp is treated with alkali and then with carbon disulphide. This yields "viscose," which can then be forced through fine jets to produce threads. Or, again, the wood-pulp may be treated with acetic anhydride, forming acetyl-cellulose which can be dissolved in acetone and driven through the thread-forming machinery.

Readers who can recall the last decade of the nineteenth century and compare it with present-day affairs will appreciate the change which has come over the scene, especially in women's attire. At the beginning of this century, a pair of silk stockings was an almost unthinkable luxury for a woman of the lower classes. But the production of rayon has brought a vast change; for, at a little distance, rayon and silk look much the same, so that now-adays it is possible for a woman to dress inexpensively and yet look, for all practical purposes, as well-attired as her richer sister, with her more expensive fabrics, especially with the simpler fashions now in vogue.

There is still, of course, the differentiation due to taste. But taste can be developed in course of time, and already the two previously-separated classes are tending to approximate in their outward guise. Further, in addition to the mere superficial aspect of clothes, there is another effect which rayon and its likes are producing. Psychologically, a well-dressed individual has an advantage over an ill-dressed one; and the leveling-up of the dress-standard is bringing its influence into social affairs to a very marked extent, as any observer can note. It is too soon yet to gauge the ultimate effect of this; but it would be mere blindness to deny its existence.

The extent to which chemistry now permeates daily life can be gauged by a very simple review of the materials which are used in clothing by the first man in the street. The leather of his shoes has been tanned by a chemical process, and the nails in the heels owe their existence to metallurgy. His suit has been dyed, so has his neck-tie. His underclothing has undergone bleaching and possibly dyeing as well, and his collar also. Any metal objects which he wears, such as a belt-buckle, studs, sleeve-links, and the like, owe their existence to metallurgy. Even if he uses bone buttons or studs, they have been shaped by metal tools in the making of which chemistry played its part. If he carries an umbrella, its metal frame was produced by metallurgy and its fabric was dyed and rainproofed by chemical methods. His hat, if he wears one, has also been dyed, and the inner band of it has been tanned by chemical processes.

In the foregoing pages only a few instances out of a multitude have been chosen to illustrate the influence which chemical knowledge and methods have exerted upon our social structure. They could be multiplied to an indefinite extent had space been available. But sufficient has already been said to indicate that this field could hardly be ignored in any history which claimed to be well balanced. Certainly the intervention of chemistry in human affairs has been of vastly greater import than, say, the abortive rebellion of Jack Cade, the Monmouth rising, the corruption of Lord Verulam, or the fiasco of the Old Pretender, which have not been overlooked by chroniclers. It seems time that chemistry, and indeed science as a whole, was given its due share among our records of social development. I present this idea to the next writer who proposes to compile a Universal History.

XI
NOVELTIES IN THEIR DAY

TRULY, THERE ARE FEW more desirable possessions than a well-stocked memory. Why venture forth to a cinema through night and wind and rain, when one can sit in an easy chair at home, pleasantly relaxed after the work of the day, and watch on the smoke wreaths that wonderful documentary film which runs without effort through "memory's gratis biograph"? The programme can be chosen to suit one's mood of the moment; and if a picture does not please, it can be replaced by something more attractive, without even the trouble of turning a dial. Switch on the instrument, then, and see what comes up.

I was born in the late Victorian era. It is the fashion in some quarters, nowadays, to speak slightingly of the old Queen; but towards the end of her reign she undoubtedly suited the times in which she lived, and what more can one ask from a representative figure? But in those days there was a mightier even than Victoria: the great god Thrift. Saving had added itself to the Seven Virtues and had been content with no modest place in the hierarchy. Our elders strove—unsuccessfully in my case, I fear—to inculcate a habit of thrift into my own generation; and in most houses one came across that abhorred device: a child's savings bank. Sometimes it was merely a gaily-painted box with a slit for inserting coins; but often mechanical ingenuity was called in to adorn the little one's path towards economy. I remember two or three of these insidious affairs.

One of them was cast in metal and represented a horse and rider, facing a hollow tree stump in the trunk of which was a cavity to receive the coin wrung from one's pocket money. To operate it, a spring was compressed; the coin was placed in the rider's hand; then, on touching a trigger, the horse flung up, its heels; the rider was thrown forward; the coin slipped from his hand into the appointed place and rattled down into the base of the bank, whence it could be extricated again only by the use of a key which was retained in parental charge. Surely a most enticing manner of saving money.

Another of the wretched things was in the form of a negro with extended palm, on which the coin was deposited; then, by pushing a lever behind the image's shoulder, it raised its hand, rolled its eyes gloatingly, opened its mouth, and engulfed the money.

These affairs, by their antics, deadened the pangs of saving; but there was another type which had not even the grace to hide its naked avariciousness. It was made of china, in the form of some animal: a cow, a pig, or what not; and it had a slit in its back for the insertion of coins. It was an especially loathsome invention, for it had no key like the others, and had to be stuffed with money before one was permitted to rescue one's little hoard by breaking the accursed thing to pieces. At least, that was the theory. In practice one might surreptitiously, and with much expenditure of time and energy, coax a coin or two back through the slit.

Nor was it only pocket money which vanished into the maws of these rapacious affairs. One ran a considerable risk of losing a part of those godsends to impecunious youth: tips from visiting relatives with generous instincts. But at least the older generation practised what it preached. People shot their savings into the big kicking horse banks of the Funds and the limited liability companies. But they seemed to like doing that, whereas I never met a child so miserly as to enjoy patronising his own savings bank, whatever its form.

These things seem unknown to-day. Certainly I never come across them, even in old country houses where ancestral Toby jugs and suchlike objects of art and virtu are piously preserved. Possibly the present generation has exasperating memories of them and refuses to give them storage room. Times have changed.

In those days, we were easily pleased. For a child with an eye for colour, the kaleidoscope was a toy furnishing endless interest. Give the thing a shake or a twist, and one dazzling polychromatic pattern vanished, yielding place to a fresh one. And we were content to turn the zoetrope or "Wheel of Life," and watch the incessant jumping of a dog over a stile or the jerky evolutions of an acrobat endlessly repeating his manoeuvres, time and again.

No Christmas party was complete without a magic lantern. At times—wild extravagance!—there might even be twin lanterns for the production of dissolving views. The slides varied: one never knew what would turn up. It might be the Well of St. Keyne, with somebody reading out the accompanying verse in a very hearty voice; or, again, Little Red Riding Hood in brilliant colours with a perfectly horrible wolf; or, again, Curfew Shall Not Ring To-night, including a slide showing a damsel (with her skirts decorously unruffled) clinging to the clapper of the bell and swung out over the abyss. And there were rackwork slides, too: educational ones showing the Earth, with a Moon of about equal size revolving solemnly around it at an absurdly short distance and exhibiting phases in its circuit; or else "funny ones" like that displaying a nightcapped man in bed, snoring with his mouth open, whilst an interminable procession of disgusting rats advanced from the foot of the bed and vanished, one by one, into his open jaws. This last kind, I confess, roused much more enthusiasm than the educational type.

If the magic lantern pleased us, it seemed equally popular with the young men and girls who had been invited "to help to look after the children at supper." Doubtless the darkness had its advantages for the engaged couples, the nearly-engaged, and probably for others as well; and the soft hissing of the limelight oxygen jet covered whispered conversation of an intimate sort. I cannot speak from experience; for by the time I had grown up to a suitable age, the magic lantern had passed into limbo, so far as I know. Still, it must have been acceptable in its time.

It was about this period, I think, that I learned of the existence of the Manx Mannikins. Advertisements appeared broadcast in the newspapers under the heading:

THE MANX MANNIKINS
Living Heads Without Bodies

Then followed a description of these surprising creatures. They had been discovered, it seemed, in some caves in the Isle of Man. They were warm-blooded, would wince if hurt, bled if one pricked them, and by their facial contortions they could express emotions which they could not voice, for they had no power of speech. They were very much attached to their human owners. Professor Huxley, the great biologist, was the possessor of a remarkably fine pair of them.

What properly inquisitive child could resist such a lure? I pestered my father to give me one of these extraordinary pets, which could be obtained at a moderate cost. He laughed, consented, and wrote for one of them, carefully giving my name and address so that I might have the pleasure of getting the parcel myself. Probably he, also, was curious to know how the advertiser could justify the terms of his announcement. For the next night or two, I went to bed with my mind full of my Manx Mannikin, trying to picture its exact appearance, hoping that I got a good-tempered one, enjoying all the prospective delight of showing this marvelous pet to my young friends. In fact, I think I got full value for the cost of the thing even before it arrived. Such is the power of imagination. Never did the time between posts seem so long.

At last it came: a little packet about the size of two 10-cigarette cartons. The Mannikin must be very small; I had imagined something rather bigger than it seemed to be. Great excitement! The wrappings were torn off with trembling fingers. What would it look like? Then, with the lifting of the box lid, perplexity. Surely there had been some mistake, a mix-up between two orders. These queer little wax objects could not be my Manx Mannikin.

They were, however, all the Manx Mannikin I was ever to receive. My father got a second laugh, his usual kindly one, at my expense. While I stared in dismay at the contents of the box, he had been reading the accompanying circular; and now he intervened to soothe my disappointment. He was very clever with his hands. He picked up the little wax objects, turned his back for a moment or two, then faced me again—and there was my Manx Mannikin almost as I had imagined it: a black-eyed little face, a rather blobby nose, a red tongue that moved, and a jaw that champed. "It's clever, isn't it? You can puzzle your friends with that." And so my regrets were displaced by anticipations of future amusement.

Of course the reader will already have seen through the trick. The Manx Mannikins were simply human hands with the addition of the little wax noses, eyes, and tongues. One clenched one's fist, fitted a pair of eyes on stalks into the gap between the forefinger and middle finger, bridged the gap between the middle and third fingers with a waxen nose, thrust a model tongue between third and little fingers, and then surrounded the whole affair with one's handkerchief to conceal the rest of one's hand. A slight movement of one's digits sufficed to produce various expressions on the little face. When I had mastered the management of the thing, I was almost—but not quite—as pleased as I would have been with a living Manx Mannikin; and in a dim light the thing was quite able to hoax many people completely. And if the reader will look back at the contents of the advertisement, he will see how slightly it diverged from the truth.

Older heads than mine were taken in by that ingenious advertisement. In the *Life* of Professor T. H. Huxley* one finds that it caused him some amusement. Writing to his wife, who was away from home on a visit, he mentions one result.

> "Yesterday two ladies called to know if they could see the Manx Mannikins. I think of having a board put up to say that in the absence of the Proprietress the show is closed."

Such was the penalty for having "a remarkably fine pair of"—hands.

---

* *Life and Letters of Thomas Henry Huxley* by his son, Leonard Huxley, Eversley Edition, Vol. III, p. 283.

Those were the days when each monthly part of the *Boy's Own Paper* bore on its brown cover the fascinating advertisements of Theobald & Co. of blessed memory. They were the first things one turned to after having read the serials to learn how the various heroes got out of the scrapes in which they had been left at the ends of the previous month's instalments. What a shop that was! Model steamboats, model locomotives, mechanical toys, telescopes, actor's make-up sets, conjuring tricks, boxes of games, puzzles, novelties of every sort that the heart could desire. One suspects that Little Lord Fauntleroy's grandfather, the Earl, did his shopping at Theobald's when he amassed that amazing collection of playthings to divert his grandson. At the cost of a five-pound note—and what was a mere "fiver" to one so affluent as he?—he could have kept Lord Fauntleroy entranced for weeks; for children were not hard to please in those days, when "Pigs in Clover" would keep a pertinacious youngster amused for hours.

Sometimes I wonder if the modern sharp-witted child does not pay as heavily for his acumen as Adam did when he ate of the Tree of Knowledge. He loses those simpler joys which satisfied his forefathers. The Boy's Box of Conjuring Tricks would not puzzle him for a moment. Show him the Magic Bottle which stands on end or lies flat at the will of its owner, and he says contemptuously: "You put something into it!" which is true, but disconcerting to the demonstrator. He fathoms the secret of the Obedient Ball, which slides down a piece of twine and stops at the word of command. "You pull the string tight when you want it to stop. Anyone can see that!" Even the card with two pips on it, which appears as an ace or a trey according to the way you hold it, will hardly pass muster twice with him. And so Wonder vanishes from the nursery.

I miss from the toyshop windows those boxes of Parlour Pyrotechnics which were so popular when I was young. Here, I think, there should be profit in a revival, for there is no child too wise to resist the temptation of playing with fire. There was the Lightning Paper which went off with a gentle "Fuff!" and a tremendous flash when you kindled it. There was the Fern Paper which smouldered away and left a beautiful frond-like ash. And, of course, there were Fire Drawings. You put the glowing tip of a blown-out match to the asterisk on the paper, and a spark began to travel over the sheet burning an outline, until finally an excellently-sketched pig fell out. Or else you found a picture of a sportsman aiming at a hare and applied the same encouragement; whereupon a spark ran out from the shot-gun, and when it reached the quarry the unfortunate hare exploded owing to the percussion cap glued on the back of the sheet. All this in addition to miniature star-lights, Bengal lights, and other amusing affairs.

Then there was the Boy's Chemical Cabinet, which enabled one to show one's friends a glass of clear water and then to turn the liquid into wine, merely by stirring it with a glass rod: the Miracle of Cana on one's own hearth rug! And one could even go one better; for by giving another stir, the wine changed back into water again for the benefit of members of the Young Abstainers' Union. The boy of to-day is not to be impressed by such affairs; he has had some chemistry at school and worked with indicators himself. Phenolphthalein is no mystery to him, as it was to us at his age. But if knowledge spreads in one direction, it sometimes recedes in other fields. The eggs of Pharaoh's Serpents were familiar enough to the children in my young days; but in demonstrating the properties of mercuric thiocyanate to my First Year classes I have been surprised to note how many of the students had never seen a Pharaoh's Serpent before. So there is, perhaps, something to be said for the old Boy's Chemical Cabinets. One of them certainly served to arouse my interest in chemistry and, under my father's skilful but unperceived guidance, turned me eventually into the path of chemical research along which I have wandered with pleasure for more decades than I care to recall, reaping the sort of scientific "European reputation" which has been described so caustically—and alas! so accurately—by Mr. Edward Shanks.*

---

* "Jeremy Tuft . . . had published a series of papers on the Viscosity of Liquids, which had gained him a European reputation—that is to say, it had been quoted with approval by two Germans and a Pole, while the conclusions had been appropriated without acknowledgement by a Norwegian."— *The People of the Ruins*, by Edward Shanks, p. 7.

Nowadays I meet many people who lightly describe themselves as amateur photographers; but what they really mean is that they snap the shutter and let somebody else do the dirty work. Amateurs, forsooth! How little do these butterflies know of the stern joys of the art as it was practised before the advent of the roll film and daylight loading. I pity the poor things—though nowadays I use a Kodak like the meanest of them—for I went through the mill in my early days and enjoyed every stage of the process from filling the camera down to peeling the finished print off a ferrotype plate.

My first steps in photography were taken with a Demon camera belonging to my elder brother. No one could call it expensive: "Camera Only, 5s.; Complete Outfit, 10s." It looked rather like a squat funnel resting on a thin rectangular base; and it used plates about two inches square, which were carried in flat, light-proof bags, since the Demon economically disdained dark slides. There were, I think, three of these bags in the Complete Outfit; and they could, of course, be reloaded in comfort after dark. But what excited beginner—especially a small boy—would be content with three snapshots a day?

Somehow or other, daylight loading of the bags was essential; and no dark room was available. There is always a way, however. One grasped the bags and a packet of plates, and one crawled between the sheets of one's bed. In this dark and stuffy hermitage, the wrappings were torn off the plates, the bags were loaded, and the remaining plates sealed up, after a prolonged search among the bed-clothes for the necessary wrappings. Often, in the heat and hurry, one forgot which was the emulsion side of the plate; and it got into the bag wrong side foremost. Often, too, the rubber strap which acted as the bag's final seal went astray and had to be groped for. Never had a bed seemed so vast and complex as during the search for one of these elusive straps. Sometimes they were so elusive that one had to abandon them and chance it. Then, freely perspiring, half-stifled, but triumphant, one emerged once more into the light of day.

After the shutter-snapping was over, the serious part of the business began. When dusk had deepened sufficiently, one pulled down the blinds, learnedly mixed one's solutions, lit the guttering candle in the little red lantern of the Complete Outfit, and turned out the gas. Out of its bag came the little plate with its stainless creamy film; into the developing dish with it; and swish! the developer swept across it. Now, what would come out? For by this time the bags had got completely mixed up and one had no idea which picture was under treatment.

Sometimes a grey veil crept uniformly over the creamy surface. Fogged! Something had gone wrong during the daylight loading process. But more often one saw dark patches sharply defined on the cream. What's this? Must be the sky; the plate's wrong side up; turn the dish and have another look. Yes, that's the sky; and here is the summer-house coming out; that black-trousered nigger in the corner must be Jack—he was in flannels. And so on, as the details emerged one by one. Finally, a glance at the back of the plate to see if it was "done," a rinse under the tap, and then the negative went into the fixing-bath. After that, an hour's washing in running water, and that stage was over, except for the drying of the negative.

The artist is fortunate who attains to beauty without pain. Those were the days when pyrogallic acid reigned as a developer; and it dyed one's fingers a dark, reddish-brown, displeasing colour. One could get rid of some of the stain with pumice-stone if one was prepared to remove one's epidermis as well; but the developer soaked into the quicks of one's nails, from which nothing would budge it. Hence maternal horror, paternal remarks about the advantages of using soap and water before appearing at table, fraternal jeers. What did such things matter, with that beautiful negative achieved? And there were still joys to come.

In those days there were no Gaslight papers: P.O.P. (printing-out paper) was what we used. One clipped the sensitive paper behind the negative in the printing frame, stuck it up at the window, and watched over it in a state of fidgets, opening the printing frame every few minutes in one's anxiety to be sure that over-printing did not occur. Then, in a shady corner, one could take the print out of the frame and examine the whole picture for the first time.

After that came the joys of toning, fixing, squeegeeing, and finally, peeling the print off the ferrotype plate which had given it that brilliant polish so much esteemed in those days. Sometimes it stuck and tore, and one lost all the fruits of one's labour in the last stage. Uncertainty dogged one up to the end.

The Demon was succeeded by a "detective" box camera about the size of a shoe box. Why it was called a "detective camera" no one could ever tell; for if it did not look like a camera, it certainly looked like nothing commonly carried by man, and would have betrayed immediately any detective who used it. It used quarter-plates held in metal sheaths. In operating it, one took the photograph; then one pulled out a stud, and was rewarded by what may be described as a loud clanging thud. This was due to the exposed plate falling film downwards on to a platform. One then pointed the lens to the sky, whereupon the plate and sheath slid through a slit and fell with a bang into a rear chamber within the box. As these clangs and clashes could be heard at a considerable distance from the operator, there was not much secrecy about the business. Nor were the results altogether beyond criticism. Every speck of dust gathered by the sensitive film resulted in a "pinhole" in the final negative; and in the course of all this clashing and clattering it collected them "more than somewhat," as Mr. Damon Runyon would say. If one took a landscape, its sky generally appeared dotted with black meteorites caught in transit; whilst if one took a portrait, the sitter was apt to seem suffering from measles or smallpox. And, by some chance, against all probabilities, a black beauty-spot usually appeared on the tip of a girl's nose if she allowed herself to be photographed. But people liked having their photographs taken in those days, and would even treasure the results, beauty-spots and all.

With the spread of the pastime, a new set of conversational topics grew up among the initiate; learned discussions about how hydroquinone yielded a "pretty" negative with strong contrasts; rodinal and eikonogen gave "thin" negatives; the abhorrent "pyro" yielded a dirty-green negative which printed best of all, and so forth. And when this subject had been thrashed out to exhaustion, one could turn to argue the relative merits of gelatino-chloride, collodio-chloride, platino-matte and ferroprussiate printing papers and wax eloquent on the values of toning with gold, platinum and uranium; or even pass on to the artistic effects to be obtained from the carbon process. Not since the days of the alchemists had such a flood of esoteric terms flowed into conversation to the amazement of the uninitiate. What does the modern shutter-snapper know of such delights?

But photography was not the only novelty in those times. I remember paying sixpence—and sixpence was sixpence to me in those days—to make personal acquaintance with another neoteric contrivance. One was led to a table whereon stood a box from which branched tubes like those of a modern stethoscope. Under direction, I jammed the ends of a pair into my ears and the attendant touched a trigger. Then began some grinding and scratching, which seemed a poor return for my sixpence. But suddenly and maliciously the machine squeaked in my ear: "I am Edison . . ." and proceeded to tell me all about itself. A generation which has grown up familiar with the radio-gramophone will hardly appreciate the shock. It was one of the early phonographs, rarities in those days. One had expected it to mumble something impersonal, recite poetry or declaim some well-known passage; but one was quite unprepared for this direct personal appeal. Edison's own voice! Marvelous! It really did speak, after all! One had been faintly sceptical about that until one heard it.

Then there was another marvel which came, went, and was never heard of again, so far as I was concerned: Oboth. I paid my sixpence—it was the usual charge for things in the novelty line—and found before me a glass globe, half-a-yard or so in diameter, filled with water and supported on two glass pillars fixed to a table covered with a cloth which drooped to the ground. A courteous demonstrator extracted from a glass of water a carrot-shaped rubber figure rather like my idea of a mandrake; and with a fountain-pen filler he blew something into the creature's noddle. Then he dropped it into the big glass globe through an orifice at the summit, and it remained floating upright at the surface of the water.

"Just tell it what you want it to do," said the attendant. "Oboth! Go down to the bottom!"

Immediately the thing sank like a stone to the bottom of the sphere. I tried it with various evolutions. "Make it lie horizontally, please." Hardly were the words out of my mouth than the thing obeyed. "Now make it go to the left-hand side of the globe; make it spin; make it lie slant-wise . . ." and so forth. Sometimes when the meaning was plain, Oboth would have finished its movement before I had ended my sentence. Finally, after I had what I regarded as a good sixpenny-worth, the demonstrator brought Oboth to the top of the sphere, extracted it, and restored it to its glass, uttering the pleasant hope that I had been interested. No! He made no claim to have discovered any new Force. There was the thing; take it or leave it.

Obviously there was a confederate within range of my voice who manipulated the controls. Magnetism was what one suspected; probably a pivoted electro-magnet under the table with the confederate to manage it; and no doubt the little rubber figure had a couple of bar-magnets inside it at right-angles to each other. It certainly puzzled older and sharper brains than mine in those days. A distinguished Professor of Physiology, I know, spent a large number of sixpences in trying to solve the mystery of Oboth; but like Omar Khayyam, came out by the same door wherein he went. And yet he made a hobby of novelties and the solution of such puzzles. He did manage to discover certain evolutions which the thing could not execute, but that was all.

Oboth left me with nothing but an entertaining and rather puzzling recollection; but there was a pavement toy in those days which not only amused me but proved useful in my Honours lectures two or three decades later. It was called the Kangaroo Balls, and as it is a pretty toy, not difficult to construct, I may describe it for the benefit of parents who possess a lathe. It consisted simply of two spheres, about the diameter of a ping-pong ball, turned from light wood. These were joined together by a single strand of the kind of rubber now used as a motive power for model aeroplanes, the ends of the strand being fixed firmly into holes in the spheres and held by driving in wooden pegs which were then cut off level with the surface. To start operations, the elastic was wound up, as in a model 'plane, until the two balls were brought close together. This could be done, once one had caught the knack of it, by holding one ball in one's hand and making the other one spin round in a circle in contact with the floor.

Once the rubber had been twisted, one released the ball from one's hand, whereupon the arrangement began to spin on the pavement with considerable velocity, the two balls rotating about a common centre like the two components of a double star, and gradually retreating from each other as the rubber untwisted. By the time all the twists were out of the strand, the affair had attained high speed and it over-ran its dead point, so that the rubber began to twist up in the opposite direction and the balls to approach each other again. Then once more it would over-run its dead point, reverse, and unwind itself again. And so it went on for quite a while, making a rather pretty affair to watch. Of course it needed a smooth surface to run on; it was not a success on a rough carpet, since the friction was too great for the rubber to overcome.

A couple of decades and more went past and I had almost forgotten that old toy. Then came the application of the Planck Quantum Theory to molecular spectra, and the picture of the molecule of hydrogen as composed of two heavy spheres approaching and receding from each other whilst the whole system spins about its centre of gravity. I wanted a model to show to my students; for a thing seen stays better in the memory than a thing which is merely talked about. Something after the style of the "governor" used in steam engines might have filled the bill; but it had too many joints and levers. A lecture demonstration model has to be simplified to its very bones if it is to be effective. And then I remembered the Kangaroo Balls. As the reader will see, they fitted the requirements precisely. So even the toys of long ago may find new usefulness if one remembers them at the right moment.

There were various optical toys, too, which came into my hands and gilded many of the passing hours. The first which I remember was a pocket microscope. It was simply a tin case about four inches long, and its lens boasted no achromatic refinements, since it was merely a bead of glass; but at least it enabled one to study the habits of cheese-mites, paste-eels, and the infusoria from the nearest stagnant pool. By its magic aid one could set up a miniature Zoo on one's own premises at no greater cost than dipping a bottle into a pond; and one certainly learned that life can manifest itself in the strangest forms: Gorgon and Hydras and Chimaeras dire. "Good Heavens! If I were reduced to the size of these things and put amongst them, what chance would I stand?" A sobering thought for human pride. I progressed later to more complex instruments, but I doubt if they brought me so much pleasure as I got from that crude little toy, since it gave me my first glimpse into the world of the invisibly small.

Then some kind friend presented me with an old-fashioned polariscope. Scatter a few fragments of mica on the inspection plate and one saw colours of marvelous brilliance and purity which fascinated the eye. No artist, not even the masters of stained glass who beautified our cathedrals, could match these hues. And there was the charm of the first impression from that old contrivance, lending something to the tints which I never recaptured in later days in handling the most elaborate polarising microscopes in the examination of rock-sections.

Astronomy came to me through the medium of a pocket one-inch telescope given to me by my father. I suppose it cost only a few shillings then, but it was a workmanlike little thing, and I have it still. With it in hand and with the help of *Astronomy with an Opera Glass* and Fowler's *Popular Telescopic Astronomy*, I groped my way about the heavens in search of double stars, nebula, the satellites of Jupiter, the lunar craters, and the spots on the sun. Amazing what amusement could be bought for a small sum in those days!

After the pocket telescope came one of two-inch aperture on an altazimuth stand—later converted into a crude equatorial mounting at no greater cost than a piece of twine—and hours spent in the garden on winter nights when the "seeing" was good. One's body might shiver in the frost, but one had an internal fire of enthusiasm to balance that. By a piece of good luck, one of the first things I saw with the new instrument was Saturn with its ring, which chanced to be in a specially good aspect for observation. One had seen pictures of it in books long before; but they were dull things compared to this golden-girdled orb floating yonder in the immensity of space. Nothing like seeing a thing with one's own eyes.

But even a two-inch aperture will reveal more things than one has time to look for in the skies: the tiny globule of Mercury glimpsed in the dawn or in still evening; the silver crescent of Venus—"just like the Moon!"; Jupiter with its belts, its enigmatic Red Spot, and its ever-shifting moons and their eclipses.

Not long after this, a friend kindly lent me for a short time a 3½ inch telescope. At that time the problem of the Martian canals was very much to the fore, and I longed to see one for myself. Many other observers had seen them with instruments of this size; why not I? But I fear that I have not the right kind of imagination for this work; and I must confess that neither prayer nor fasting would have helped much in the matter, for even an instrument of thrice the aperture failed to make me personally acquainted with these elusive objects. Probably my eye is not sufficiently sensitive for the task.

When I came to put in an Astronomy course for my degree, I was given the freedom of the big Ochtertyre refractor under its dome on Dowanhill. But by that time the early thrills of star-gazing had faded; and although I enjoyed the hours I spent up there with only the slow-ticking sidereal clock for company, they never gave me anything like the pleasure I got from that first glimpse of Saturn's rings with my own little two-inch instrument.

Talking of old toys, I am reminded of Professor Augustus de Morgan's *Budget of Paradoxes** which charmed me in my younger days, and to which I still turn at times for amusement. In his library, de Morgan had a vast collection of books and pamphlets written by gentlemen who wished to prove the flatness of the Earth

---

* *A Budget of Paradoxes*, by Augustus de Morgan, London, 1872.

or who believed that they had found means of squaring the circle, duplicating the cube, or trisecting an angle by geometrical methods; and in his *Budget* he reviewed some of these masterpieces. Occasionally he diverged into other matters, and in one of these excursions he reviewed a pamphlet entitled: *The theory of the Whizgig considered; in as much as it mechanically exemplifies the three working properties of nature; which are now set forth under the guise of this toy, for children of all ages.* London, 1822, pp. 24.

Here is part of what de Morgan has to say about it. "The toy called the *Whizgig* will be remembered by many. The writer"—of the pamphlet—"is a follower of Jacob Behmen. . . . Jacob Behmen first announced the three working properties of nature. These laws are illustrated in the whizgig. There is the harsh astringent, attractive compression; the bitter compunction, repulsive expansion; and the stinging anguish, duplex motion. I do not think (remarks de Morgan) that this mention will revive Behmen; but it may the *whizgig*, a very pretty toy, and philosophical withal, for few of those who used it could explain it." What was this now-forgotten *whizgig*? It sounds interesting, and one would like to learn more about its nature and activities.

The gullible were with us, *consule Planco*, as they are to-day. I remember a modest example: the Harness Belt, which was intended to appeal to sufferers from rheumatism. It was a broad leather belt studded with discs of zinc and of copper arranged alternately on its surface; and when the rheumatic patient wore this up-to-date substitute for a hair shirt next to his skin with faith, electrical currents were supposed to be set up on the epidermis by the metal studs acting as the electrodes of a battery with, presumably, the perspiration serving as an electrolyte. How it came about, I cannot recall, but eventually the Harness Belt produced a law case as a by-product of its other activities, and Lord Kelvin was called as an expert witness. "Was there," he was asked, "any electricity in this contrivance?" "As much as there is in an empty pill-box," was the devastating reply.

I was much impressed when my father read this out from the newspaper at the time; for Lord Kelvin—Sir William Thomson as he then was—had been a familiar figure to me almost since my earliest days, and I had difficulty in picturing that genial, kindly and sympathetic personality destroying the opponents' case in this unceremonious fashion.

My father's friendship with Kelvin long predated my arrival in this world, for he had been one of Kelvin's students, as I was later to become myself. From time to time, during my childhood, I paid visits to the house of Sir William; but I fear that my interest was not so much in the engineer of the first Atlantic cable as in his parrot, "The Doctor." That was a bird with an unmistakable sense of humour of the most wicked kind. It was an amazingly good talker, and could imitate its owner's voice to perfection; and it used this faculty without remorse. Sir William had a pet dog, Jack. Through the house would echo the call: "Jack! Jack!" and the wretched dog would run in joyful obedience to greet its master. All it found, on rushing into the room, was "The Doctor," who would survey it sardonically with an air which said plainly "Aha! Caught you again, my friend! Will you *never* learn?" Then with a gesture of manifest contempt, it would bend down and whet its beak on its perch, still keeping its eye fixed on the unfortunate dog, which shamefacedly slunk out of the room on discovering how it had been deceived.

Kelvin, in earlier years, was a pent-up store not only of mental but of physical energy as well. "When I was a student of his," my father told me, "Kelvin was one of the handsomest men I ever saw. I never saw him walk through the laboratory; he always ran." But when I knew him, this had passed away. One day, when out curling, he slipped on the ice and broke his thigh-bone; the local medico was called in; the fracture was badly set; and thereafter Kelvin hobbled until his death.

He had an extraordinary alertness of mind and fertility in ideas. Once a humorously-inclined student dropped a marble in his lecture room and it rolled down the tiers of the benches: "Tap . . . tap . . . tap . . ." until it emerged on the floor in front of the lecture bench. Kelvin looked at it, ran his eye up the class: "Mr. So-and-so, you dropped that marble, I think?" He had counted the taps, noted the spot where the marble emerged, and counted up the tiers—one tier for each tap—and so was able to name the joker almost before the marble had rolled a yard on the floor.

On another occasion, he had sent a couple of not very bright students to the top of the tower to measure the potential of electricity in the air. When they showed him their results, he was surprised, as they were not at all what he had anticipated. Next morning, in his lecture class, he was so engrossed by the anomalies that he spent the whole hour in pouring out, one after another, a series of hypotheses to account for these extraordinary figures. One explanation was, if I remember aright, that the engines blowing off steam in a near-by station had generated frictional electricity in the local atmosphere. Alas! the true explanation was much simpler. The two dunderheads had been reading the scale of their instrument from the wrong end!

In the days when I was a student of his he lectured only at irregular intervals, leaving one of his assistants to give the routine of the course. He would begin, always, with something well within our grasp; but almost immediately something else engrossed his attention, and before we well knew how it had come about, he would be discoursing on subjects far beyond our ken. Undoubtedly, in those days he was one of the worst lecturers I ever heard, if the object of a lecture is to convey knowledge in comprehensible form. One day, as he made an unexpected appearance, I heard the man in the bench behind me groan: "Oh, Lord! I brought my notebook this morning—and now it's Kelvin! Curse!"

Kelvin's own work called for only the higher mathematics, and he had gradually forgotten his arithmetical pothooks. Sometimes, when he came down to our level, he had to make a simple calculation on the blackboard. Then one might catch an anguished whisper addressed to the lecture-demonstrator: "Black! Black! What's six times nine? Quick!"

Long after his retirement from his Chair, he retained the keenest interest in physics and kept abreast of the most modern discoveries in that field. Once, at some meeting, he came up to my father at the close. "Come home with me; I want to show you something." And when they reached his house, he produced the "dust-counter" which had just been devised by Professor C. T. R. Wilson; and demonstrated its working with as much pleasure and enthusiasm as if it had been one of his own multitudinous inventions.

Kite-flying seems now less popular than it was when I was young. Perhaps the coming of the aeroplane killed it. But in my time we were all enthusiastic kite-fliers. I must have begun long before I reached my teens; for in those days sixpence—I seem to be mentioning sixpences continually; perhaps it is my Scots blood coming out—sixpence, I say, would procure in any toy-shop an excellent little Chinese kite of bamboo and stout tissue paper suitable for anyone above the tottering age. A reel of thread obtained from domestic sources completed one's outfit and allowed one to launch this gaily-coloured butterfly in the lightest breeze.

The bow-headed kite followed, and then, if one wanted something larger, the big hexagonal pattern. Later, the box-kite arrived, but it I never loved. It was too stable, and flying it was a dull business. Besides, it had no tail; and in the tail resides half the beauty of a proper kite. There was no sort of adventure in flying a box kite, none of those graceful swoops and dives with curving tail which constituted much of the pleasure in kite-flying. To see your kite hurtling earthwards to destruction and to know that rescue depends on your proper handling of the string; these were the moments which tested your skill. And when your toy escaped disaster by a hair's breadth and soared up into the blue once more, you felt both relief and pride in your achievement.

But there were flying contrivances other than kites. One was revived not long ago after a long lapse: a light wooden air-screw with a round shaft which one spun between one's palms and then let go. It would keep itself suspended for a few seconds. A more interesting affair was built of bamboo and tissue paper in the shape of a butterfly with broad antenna which were driven round by a twisted rubber band. When released, the thing flew erratically about a room in a most attractive manner. Then there was the helicopter toy: a four-bladed air-screw with a circular rim, lightly built of tin. To fly it, one put it on a handle, pulled a string coiled round a reel, and the helicopter flew ups spinning as it went. It would rise fifty or sixty feet in the air in a very graceful fashion. But when one set it agoing in the garden it seemed to be incapable of avoiding the house roof, on which it got lost; and then one had to buy a complete new contraption, handle included, as the manufacturers cunningly refused to sell its parts independently of each other.

Somebody has offered as a cure for minor worries: "Go and look at an elephant." It is not bad advice. The spectacle of that great grey slow-moving creature is undoubtedly soothing to the troubled mind. In my youth I discovered that a similar curative effect is produced by watching a big engine performing its tireless evolutions. The throw of a huge connecting-rod produces a tranquilising effect. I had not heard of the elephant-cure then; and in any case I had no elephant to study, within easy range. But I often passed a display window in which, as an advertisement, a big Tangye gas-engine was kept running from morning to night; and its smooth, measured beating undoubtedly served me as well as any elephant would have done. Little did I think that I had before me the forerunner of the machine from which bombs would rain down in my neighbourhood; for in those days the motor-car was unknown and the aeroplane a mere airy fancy.

Not that suggestions were lacking. Jules Verne was my favourite author at that period, and my experiences with the helicopter toy led me to be less sceptical than most readers were about the practicability of the Clipper of the Clouds. Then, suddenly, familiar hoardings bore a huge picture advertising E. D. Fawsett's *Hartmann the Anarchist:* smoke and bombs and dust, with the clock-tower of Westminster falling and the *Attila* soaring gigantic above the wreckage. I procured the book, read it in feverish haste and, helped no doubt by F. T. Jane's illustrations, I became a convert to the aeroplane. After all, one threw paper darts, and what were they but tiny aeroplanes?

Then began a series of experiments with loaded cardboard planes drifted down a long flight of stairs, with much careful measurement of the lengths of flight obtained and the effects of shifting the position of the weights on the plane. George Griffith's *Angel of the Revolution* came as a spur to flagging energies. Then, perhaps fortunately, I procured Professor S. P. Langley's monograph *Experiments in Aerodynamics*, and read all of it except the mathematics, which were beyond my small limits. (My recollection of the work is that it is mainly mathematical, so it cannot have been a lengthy study if one skipped the maths.) It discouraged me completely. What use were my miserable little experiments with cardboard planes when an eminent man, four times my age at least, with a vast turntable at his disposal and mathematics as a handmaid, had not succeeded in solving the problem? *Experiments in Aerodynamics* produced such an inferiority complex in me that I discarded my toy planes, ceased to become a nuisance to my friends by discoursing on my experiments, and returned to my old love, chemistry, without a pang. Icarus fallen, without having even soared.

However, I was not entirely finished with the pioneers of aviation; for later on I made the acquaintance of Pilcher, who, like Lilienthal, led the way in gliding. His looks suggested the good sportsman rather than the fine engineer he actually was; but there was no mistaking his enthusiasm for gliding. He very kindly showed me his machine. The frame of it was constructed by himself in the Engineering Department of the University, where he was an assistant; and the fabric of the wings, which folded up, had been sewn, I believe, by his sister. After he had shown me his glider in detail, I asked: "Are you going to put an engine on to it?" "Yes, by-and-by." Only later did I understand the boredom of his tone. Dozens and dozens of people must already have asked him that intelligent question.

At that time Pilcher was carrying out his gliding experiments in a field near Cardross; and it was there, shortly after he had shown me his machine, that he came to grief out of sheer good-nature. A number of people had gathered to see his flights that day. It was unsuitable weather, with an irregular wind blowing over his flying-ground, and he might well have declined to face the risk involved. But Pilcher could not bear the thought of disappointing his audience, some of whom had come from a distance. He took the risk and went up; a gust of wind caught his machine and capsized it; and so he came down to his death. When next I saw his glider, it was in the South Kensington Museum.

It has always puzzled me to guess why a successful glider did not make its appearance earlier in history. It did not need to await the invention of the petrol engine, for in their pioneer work the Wright Brothers added a motor only after they had accomplished many engineless flights. Nor, certainly, was the delay due to any lack of intellect in earlier days. Leonardo da Vinci, probably the finest brain which humanity has yet evolved, was a skilled engineer as well as many other things. He tried his hand at devising a glider, and, as one can see from his notebook, he had the highest hopes of success. Yet evidently his Great Bird was a total failure; and he cannot even have attained to the degree of success which Pilcher and Lilienthal achieved, or he would have been encouraged to persevere in his researches. Perhaps he clung too closely to the avian design with highly-curved wing surfaces, like those of the Lilienthal and Pilcher gliders, and neglected to test the plane model which the Wrights brought to success. But Leonardo was by no means a one-ideaed man; and from his immense versatility one would have expected to see him try more than one design before throwing up the problem in despair. Yet the Great Bird vanished abruptly, and he turned his mind to other things. It is a mystery, and perforce we must leave it unsolved.

XII
THINGS IN THE NORTH

"MYSTERY WILL LEAD MILLIONS by the nose," wrote Henry St. John in disgust, more than two centuries ago. What would he say to-day, if he woke up in this world teeming with crossword puzzles, acrostics, chess problems, detective stories, and the like? Every advance in the mechanisation of our existence seems to make humanity's call for enigmas of one sort or another ever louder and more far-reaching. The Sphinx herself would be hard put to it to meet the modern demand.

For my own part, I like a good mystery. If its solution evades me, so much the better; for I can put it aside, as a dog stores a bone, and take it out again later to gnaw once more. Luckily we have plenty of problems to ponder over. What really happened to Elizabeth Canning? How did Sir Thomas Overbury come by his death? Who poisoned Bravo? What was the truth of the Borden murders? Who was Jack the Ripper? What are we to make of Easter Island and its statues? Can we find an acceptable solution for the case of the ill-fated *Mary Celeste?* Are the planets habitable, if not by "humanity in trousers," then by some other form of intelligent being? Why do astronomers not repeat the "solar experiment" and give us something more definite in the way of evidence? What about the phenomena classed under the head of "Poltergeists"? How comes it that a "lapse of memory" seems in so many cases to coincide with a full note-case, since the patient often turns up hundreds of miles from his starting-point and traveling expenses must run to more than a few shillings?

Even when we reach the truth about some of these affairs, the psychology of the actors keeps one pondering. Take the case of the Fortsas sale.* In 1840, a printed catalogue found its way into the hands of book collectors in England, France, Belgium and Holland announcing the sale of the library of Jean Népomucéne-Auguste Pichauld, Comte de Fortsas, which was to take place in Binche on August 10. From the bibliophile's standpoint, it was a marvelous collection, for it contained no book of which a second copy existed in the world. Eager buyers arrived in Binche on the day of the sale, and wandered about the little town demanding M. Mourlon, the notary charged with the disposal of Count Fortsas's effects. But, as they learned to their dismay, there was no M. Mourlon in Binche. Worse still, there had never been a Count Fortsas. The whole affair was a laborious and ingenious hoax perpetrated by M. Renier Chalon, an antiquarian and numismatist. But this did not transpire until sixteen years later, when the printer of the "catalogue" betrayed the confidence of his client. Seldom has a fraud been so cleverly contrived, for it deluded many eminent experts on their own ground.

But what can one make of the psychology of M. Chalon? It must have cost him no little trouble to compile, print, and circulate his spurious catalogue. Granted that he derived much sardonic amusement from mingling with the assembled collectors, listening to their talk, and witnessing their dismay and consternation when they learned how they had been fooled, still one is left in doubt as to his motive. Was it a practical joke on collectors in general? Or was he paying off a grudge against one particular individual and were the rest merely swept into his trap in order to make sure of his special victim? Was it a matter of humour or of malice? None can tell now.

One might well suppose that the Fortsas episode represented the limit to which practical joking could be carried at the expense

---

* A fuller account of this, with further references, will be found in E. L. Pearson's *Books in Black and Red* (1924).

of the learned, but this would be to under-estimate the curiosities of human mentality. Scotland is supposed to be the home of dry humour, barbed and tanged, and it is to Scotland that we must go to find something even more surprising than the library of Count Fortsas.

Within the last fifty years, Scottish archaeological circles were stirred by the discovery of the Dumbuck crannog. Now Dumbuck, even half-a-century ago, was no lodge in a wilderness. It was easy of access, for it lies on Clydeside, only some fifteen miles from Glasgow. Close at hand is the exit of the Forth and Clyde Canal, on which was tested, in 1802, the first practical steamboat of the world, the *Charlotte Dundas*; nearby is Bowling, whence plied the first of the Clyde passenger steamers; not far off is Old Kilpatrick, one of the numerous birthplaces attributed to St. Patrick; and the western terminus of the Antonine Wall across Scotland lies in the vicinity. Thus the neighbourhood of Dumbuck is one fitted to attract those whose pleasure it is to muse upon times of yore and days long gone before, as it says in *The Arabian Nights*.

As for a crannog, it is defined in the dictionary as an "ancient lake-dwelling in Scotland or Ireland." There was undoubtedly a genuine crannog at Dumbuck. I have seen the remains of it myself, and I hasten to warn the reader that they are not impressive. Two or three uprights in the water are all that now remain visible, for the living-platform which rested on these props in prehistoric days has long since decayed and vanished.

Our ancestors of the caves were not tidy folk. After they had sucked the marrow out of a bone they flung the rest aside on to a refuse heap at the back of the cavern, and thus were formed those instructive piles of rubbish which we now describe as "kitchen-middens," and pore over in order to discover from the relics the cultural conditions under which the makers lived and the animals with which they were acquainted. When the cave-folk migrated to lake dwellings, they simply tipped their rubbish into the water below the platform on which they lived, and so more "kitchen-middens" accumulated. Now there is not much to be learned from the contemplation of a few poles sticking up at random from a sheet of water; but the contents of the submerged kitchen-midden can tell us a good deal. Gnawed and broken bones, discarded fish-hooks, shattered pottery, lost stone implements, shells, arrow-heads, spear-points, and fire-marked materials, all give us evidence from which we can guess something of the life led by our forefathers. So, naturally, it was on the kitchen-midden of the Dumbuck crannog that interest was concentrated after the discovery of the actual site; and the finds in it established the genuineness of the crannog beyond a doubt.

Among the rubbish extracted were certain oyster shells. There was nothing surprising in this, for shell-fish formed part of the diet of early mankind, and oyster-shells are quite frequent in kitchen-middens. But the Dumbuck oyster-shells had a peculiarity never before observed in such finds. They bore scratches, crude attempts at picture making. More surprising still, the "pictures" bore a strong resemblance to some rude pictorial efforts hitherto detected only in Australia among the primitive art-work of that distant continent. Here, indeed, was a puzzle of the highest interest. Had the Australian aborigines voyaged ten thousand miles to make a settlement—and a single settlement only—at Dumbuck? Or had the Dumbuck crannog sent out from its midst the original Australian Exploration Expedition and spread its peculiar artistic technique to the island-continent long before Magellan sailed the seas? Or, by some amazing coincidence, had Clydeside and Oceania developed independently identical cultures? Here, truly, was food for speculation.

But now someone learned in oyster lore examined these mysterious shells, and the problem grew yet more complex. For these were the shells of American "Blue Points," a type of oyster found in a bed on Long Island, N.Y., U.S.A. And so another continent was dragged into the mystery.

What is the solution? I do not know. Probably no one now alive holds the key. But one may be allowed to conjecture, without pretending to have reached the truth; and here is one possible explanation. Imagine an archeologist with a keen, if distorted, sense of humour, eager—like M. Chalon—to enjoy a Gargantuan laugh at the expense of his confrères. He discovers the Dumbuck crannog; but in the interest of his joke he refrains from speaking about it and willfully deprives himself of the kudos which he might otherwise have reaped. Instead of publishing his results, he goes out of his way to fake a series of oyster-shells, which he drops among the genuine odds and ends of the crannog kitchen-midden. Then he sits back, waiting with a spider-like patience until some other archeologist chances upon Dumbuck and independently examines the kitchen-midden in the belief that he is the first in the field. The oyster-shells with their strange devices are expiscated from the fluvial waters; a violent controversy disturbs the archeological world; and the sardonic jester, secure in his secrecy, enjoys in solitude the results of his harlequinade. Admittedly, this hypothesis merely removes the mystery from its original centre and transfers it to the strange psychology of the contriver of such a plan. But can the reader suggest anything better? I leave him to try for himself.

The key to the riddle of Dumbuck would doubtless bring pleasure to its finder; but there is another problem which, if unraveled, might bring the solver not merely intellectual satisfaction but some hard cash as well: the puzzle of the Loch Arkaig treasure. It has always surprised me to see people set out to seek gold on the Cocos Islands, or in the Caribbean, or go a-diving after Armada galleons sunk near Tobermory, when all the while, only a few miles off one of the main roads of Scotland, a hoard lies buried, awaiting a discoverer. There is just as much chance of unearthing the one as the others, so far as I can see. Why, then, go to the expense of fitting out costly expeditions, when a drive up the Caledonian Canal to Gairlochy Inn and a walk of perhaps a dozen miles will take you to your hunting ground?

When the Jacobite day came to its stormy sunset at Culloden, some of the sinews of war remained unspent. These were in charge of John Murray of Broughton, who was the Prince's Secretary. Charles was no judge of character, since he regarded Murray as one of the honestest, firmest men in the whole world. Possibly he changed his opinion later when honest Murray, armed with all the official papers of the Rebellion, gave himself up to the English and bought his own safety by handing to them documents involving almost every noted Chief in the North.

In those days, Loch Arkaig was a remote fastness. During his flight, the defeated Prince visited it twice, on one occasion having the pleasure of watching through a perspective glass the hostile militia combing the district for him, only a mile away. Thither, too, came the greatest brain and greatest scoundrel in Scotland, Lord Lovat. At seventy-eight years of age, he had himself carried on a litter through seventy miles of terrible country in order to meet the Secretary, to whom he unveiled his scheme of forming a "short service" standing army out of the Clans—the only plan which offered the slightest hope of carrying on the Rebellion with any success. The Secretary promptly asked for a written document embodying Lovat's views. But Simon Fraser knew his Murray better than did the Prince; he had a shrewd idea of the use which the Secretary might ultimately make of such a record; and he refused to put pen to paper.

It was in this wilderness, "in the wood by the lochside," that Murray buried the remaining Jacobite treasure, variously estimated at 20,000 louis d'or or seven barrels of French gold. Some of this was later dug up and sent to the Prince in France, but there was a considerable residue after this had been done. So says tradition, and Highland traditions are usually reliable, for memory lasts long in the North.

Such, then, was the Loch Arkaig treasure. If the reader desires to seek it, let him read Mr. Secretary Murray's own account of the transaction. I cannot promise that he will get the gold, but he will probably get very wet. Even one M'Culloch, usually indefatigable in his topographical researches, abstained from visiting Loch Arkaig. "It is said," he tells us, "that Loch Arkaig is a picturesque lake, though unknown.... But on this I must confess ignorance, and plead misfortune, not guilt; the flight of what never ceases anywhere to fly—time; and the fall of what seldom ceases here to fall—rain."

But let us turn to another of the enigmas of Scotland: the endowments of Coinneach Odhar Fiosaiche; Dun Kenneth, the Seer. As to the existence of second sight, that rumoured power of internal vision by which future or distant events are presented, I hold no brief. Certainly I do not possess it myself and am a pure agnostic on the subject. But honest agnosticism should not preclude an examination of the evidence; so let the evidence speak for itself and then the reader can form his own opinion on the matter without any help from me.

Dun Kenneth was born in the island of Lewis, early in the seventeenth century; but he migrated to the mainland and settled in the Mackenzie country near Brahan Castle, which lies a little to the west of the road running from Beauly to Dingwall. In some way, he had become possessed of a "white stone"—possibly of quartz—with a hole in it; and by looking through this hole he saw visions on which he based his prophecies.

One of these dealt with the construction of the Caledonian Canal, for Dun Kenneth, peering through his crystal, saw a picture of full-rigged ships sailing round the back of Tomnahurich hill, where no ship could possibly have gone in his day. Actually, the first survey for the Caledonian Canal was made by James Watt in 1773—about a century after Dun Kenneth's time—and the cutting of the Canal was not begun until 1805. But a forecast of this sort was not beyond the powers of an imaginative man in the seventeenth century. The chain of three lochs in the Great Glen—Loch Ness, Loch Oich, and Loch Lochy—might well have suggested the possibility of a trans-Scottish waterway even in Kenneth's day, without the need for anything more than shrewd foresight.

Another prophecy was to the effect that Tomnahurich would be put under lock and key, and the spirits of the dead secured within. Tomnahurich ("The Hill of the Yew Wood," sometimes interpreted as "The Hill of the Fairies") is an isolated mount about a mile from the centre of Inverness. Nowadays it is enclosed and laid out as a cemetery and park, so that its present state fits neatly with Kenneth's prophecy. Yet one can hardly believe that this forecast was based merely on an intelligent anticipation of a condition of things to come, as the Canal prediction may have been.

Some of Kenneth's prophecies have been fulfilled by inventions undreamed of in his day. Assume that he really had the gift of second sight and could see far into the future, he would have no way of describing clearly what he saw in his vision, for the underlying ideas would be completely outside the scope of his experience. He would have to make shift with words as best he could. Take this prediction, for instance: "The day will come when long strings of carriages without horses shall run between Dingwall and Inverness." Imagine that a man of the sixteenth century had been shown a modern railway train. What would he have made of it? Would he not have described it in just those very terms?

Again, he foretold that: "The day will come when fire and water shall run through all the streets and lanes of Inverness." Blood and fire one could understand; that combination might well represent a poetical way of describing hard fighting in the town. But "fire and water" cannot be twisted to mean anything of the kind. Then one thinks of gas pipes and water mains. Kenneth had no knowledge of such things. He might have pitched upon water conduits by a leap of imagination; but lighting by gas could not even have been dreamed of in his times. He would have no words to fit his vision, yet one must admit that, if vision he had, he made a recognisable effort to tell what he had seen.

A final example of Kenneth's vaticinations may be given, the strangest of them all, in which he foretold the Doom of the Seaforths. According to tradition, during the reign of Charles II, the Earl of Seaforth made a journey to Paris, leaving his wife at home. Time passed; no word came from the absent Earl; and Lady Seaforth, growing anxious, called in Dun Kenneth and asked if he could give her news of her absent lord. Nothing loath, the Seer peered through his crystal in her presence, laughed aloud, and assured the lady that her husband was not only well but was merry. The adjective roused Lady Seaforth's suspicions, and she pressed for details. Kenneth, with lamentable candour, gave a vivid description of the vision which had been revealed to him: Lord Seaforth in rich attire sitting in a splendid room and toying with a comely lady on his knee. Evidently the prophet's foresight failed him on this occasion, for had he guessed the sequel he would have bridled his tongue. Lady Seaforth, probably infuriated by jealousy, wreaked her anger on the man who had dared to laugh at the picture which he had described. Kenneth, without stay, was condemned to a horrible death.

He was led out to execution, but before he died, he uttered the most circumstantial of all his prophecies. The Seaforth family would end in sorrow. The last chief of the line would be deaf and dumb, and though he would have four sons, all of them would die before their father. And in that day there would be four great lairds—Gairloch, Chisholm, Grant, and Ramsay—who would be neighbours of the last Seaforth. And one would be buck-toothed, another would be hare-lipped, a third half-witted, and the last a stammerer. After the last Seaforth had died, the family estates would be inherited by "a white-hooded lassie from the East."

An examination of the actual circumstances in which the male line of the Seaforths became extinct shows that Kenneth foretold them with surprising accuracy. The last Seaforth, in his boyhood, had normal hearing, but after a severe attack of measles while at school, he became stone-deaf. This, by itself, was not much of an advance towards Dun Kenneth's foretold conditions, which demanded a deaf mute. But the next instalment was provided by the successive deaths of all his sons during his lifetime; and with the death of the fourth and last, Lord Seaforth relapsed into silence and, as the old phrase goes, "never spoke again." Stranger still, the physical defects predicted by Kenneth in the neighbouring lairds all appeared in this generation. Thus, every one of the first nine conditions foretold by the Brahan Seer was present. And in 1815, Lord Seaforth died, the last of his family in the male line. Now came the final happening. The family estates, for lack of a male heir, went by entail to the last Lord Seaforth's eldest daughter. And who was she? The recently-bereaved widow of Admiral Sir Samuel Hood, whose name is still commemorated in our Fleet. Thus, Kenneth's prophecy was fulfilled in all its ten conditions; for even if Lady Hood's name be rejected on the ground that the Seer would never have perpetrated a pun on the eve of his execution, there is the alternative that he referred to the special head-dress which formed part of widows' weeds in those days.

There is the evidence, and there seem to be only two possible explanations: either Dun Kenneth actually possessed second sight or else he by chance hit upon a series of conditions which were fulfilled *en bloc* and in detail, scores of years after his death. If the latter solution be accepted, the odds against such an aggregate of coincidences are so great as to be hardly calculable. The reader must make his own choice.

When he came to die, the Brahan Seer took precautions that his dangerous talisman should not be left behind him to work ill on any aspiring successor. He threw it into Loch Ussie. So if the reader yearns for second sight, he may do worse than wander on Loch Ussie's strand in the hope of discovering a white stone with a hole through it, thrown up by the waters; for Coinneach Odhar Fiosaiche foretold that whoever discovered it would inherit his own prophetic gifts. I have no intention of seeking it myself, so my readers have one possible competitor the less if they embark on the quest; and it may help them if I mention that Loch Ussie lies not far from Strathpeffer, in the parish of Fodderty.

A female counterpart to Dun Kenneth is furnished by the Lady of Lawers, one of the Breadalbane family; but her prophecies seem to have been restricted to the lands of Breadalbane. They are, it is said, written in a book, shaped like a barrel and bound with twelve iron clasps; and this "Red Book of Balloch" is supposed to be stored in the charter room of Taymouth Castle. Among other things, she prophesied that when the red cairn on Ben Lawers fell down, the Church in Scotland would be split. Actually, this cairn does seem to have fallen in 1843, in which year the Church of Scotland broke in two at the Disruption.

Much more renowned than either of these is a contemporary of Sir William Wallace, variously known as True Thomas, Thomas the Rhymer, and Thomas of Ercildoune (the modern Earlston in Berwickshire). In the Highlands, he is nicknamed "the son of a dead woman," perhaps because he owed his birth to a Caesarean operation, though a more romantic version is that a cry was heard in the mother's tomb after burial, and when the grave was opened, Thomas was discovered in the coffin.

Be that as it may, he was a real person and not a mere myth, for his name—Thomas Learmont—appears in two thirteenth century charters showing that he owned lands in Ercildoune; and a poem of his, *Sir Tristrem*, is said to have survived among the Auchinleck MSS. and was edited by Sir Walter Scott. The Russian writer Lermontov claimed descent from him.

Tradition ascribes Thomas's prophetic powers to the Queen of Faerie, who found him in his youth, day-dreaming on Huntlie Bank, and carried him off into the heart of the Eildon Hills, where he remained for seven years with her in her kingdom before he was allowed to return to the world of men. He reappeared at the Eildon Tree, a spot which he is reputed to have selected thereafter for issuing his prophecies. The actual tree itself disappeared long ago, but its site is marked by the Eildon Tree Stone, beside the road west of Newtown St. Boswells, about four miles from Earlston, close to where the road used to be crossed by the Bogle Burn, that haunted brook which is supposed to have gained its name from the visits of Thomas's supernatural familiars.

Three of the Rhymer's prophecies will give an idea of his quality. "The sheep's skull will make the plough useless." This may be interpreted as foreshadowing the time when sheep-farming ousted agriculture in Scotland. "The south sea will come upon the north sea" might be regarded as a forecast of the cutting of the Caledonian Canal. "Scotland will be in white bands, and a lump of gold will be at the bottom of every glen." One may, if one chooses, read this as referring to the extension of highways through Scotland (in pre-tarmac days) and to the saving character of the Scot.

Thomas came to a mysterious end, for, like the Baker in the *Hunting of the Snark*, he "softly and suddenly vanished away." It came about in this way. When he was released by the Queen of Fairyland, she took his promise that he would return to her when she sent him a message. One day, as Thomas was making merry with his friends in the tower of Ercildoune, someone came running in; and in fear and astonishment declared that a hart and a hind had left the neighbouring forest and were slowly and composedly pacing up and down the village street. This was the expected signal. The Rhymer instantly arose, left his dwelling, and followed the wonderful animals into the forest, whence he was never seen to return; and according to legend he is still in Elfland. One version puts his abode in Dumbuck Hill, another places him in Tomnahurich.

He has not, however, completely vanished from human ken. Sir Walter Scott once began to write a tale which "should turn upon a fine legend of superstition which is current in the part of the borders where he had his residence." The tale was never finished, but Scott gives some details. A rather rumbustious horse-coper was riding over Bowden Moor, to the west of the Eildon Hills one moonlight night, when he encountered a venerable stranger in antique dress who began to chaffer for the horses which the dealer had with him. When the bargain was struck, the buyer paid in gold, but the coins were unicorns, bonnet-pieces, and similar ancient currency. He also arranged, that the coper should bring other horses for sale, stipulating that he should come by night and alone. After several horses had been thus sold, the coper broadly hinted that dry bargains were unlucky and that, as the stranger must live in the neighbourhood, a half-mutchkin would not come amiss.

"You may see my dwelling if you will," replied the stranger, "but if you lose courage at what you see there, you will rue it all your life."

Nothing daunted, the coper tied up his horse and followed the stranger up a path which led to a curious natural formation called the Lucken Hare, famous for being a rendezvous of witches. Here they entered a passage or cavern which the coper had never before seen, though he was well acquainted with the spot.

"You may still return," said the leader, ominously.

Shrinking from showing the white feather, the coper followed on and entered a very long range of stables. In every stall stood a coal-black horse and by each horse lay a knight in black armour, with a sword beside him: all silent and still as though carved from black marble. Torches burned, and by their light the coper and his companion came at length to the upper end of the vast hall where, on a table, lay a sword and a horn.

"He that shall sound that horn and draw that sword," said the stranger, "shall, if his heart fail him not, be king over all broad Britain. So speaks the tongue that cannot lie.* But all depends on courage, and much on your taking the sword or the horn first."

Fearing that the drawing of the sword might appear like a gesture of defiance, the coper picked up the horn and blew feebly on it. Immediately thunder rolled; the horses came to life, snorting, champing their bits, and tossing their heads; the knights sprang up, brandishing their swords; and the whole army seemed about to fall upon the disturber. The coper dropped the horn and attempted to grasp the sword; but at the same moment he heard a voice:—

> "Woe to the coward, that ever he was born,
> Who did not draw the sword before he blew the horn."

A whirlwind blew through the hall and carried the unfortunate coper out of the cavern on to a heap of loose stones, where he was found by shepherds next morning with just breath enough to tell his tale before he died.

Another legend explains why True Thomas required horses. He is, according to it, one of the great Un-dead like Charlemagne, Arthur, and Barbarossa; and when his hour strikes he will return to earth as a leader in some great convulsion. For that day, he buys horses, and his string is now almost complete. Each horse must have special characteristics: one must be a yellow foal with a white forehead, another must come from the meadow of Kengharair in Mull, another must be a white horse that has got "three March, three May, and three August months in its mother's milk."

When he has completed his purchases, he will become visible, and a very disturbing time will ensue.

---

* This sentence indicated that he was the Rhymer, as it was a well-known description of him.

> "When Thomas comes with his horses,
>   The day of spoils will be on the Clyde."

A great battle will be fought, and it will be possible to walk across the Clyde on men's bodies. On a particular stone in the river a bird will perch and drink its fill of blood without bending its head.* The miller of Partick Mill will grind for two hours with blood instead of water. A new king will be set on the throne. Then, with a certain air of anti-climax, sixteen ladies will follow after one lame tailor.†

It is safer to leave such legends aside and keep to facts. In the parish church wall of Earlston there is embedded a stone—the comparatively modern substitute for a more ancient one—which bears the inscription: "Auld Rymr Race Lyes in this place." A fragment of the Rhymer's Tower still stands at the south end of the town. Over a century ago, a curious incident occurred in connection with this. The veneration paid to the Rhymer's dwelling-place "even attached itself in some degree to a person who . . . chose to set up his residence in the ruins of Learmont's tower. The name of this man was Murray, a kind of herbalist; who, by dint of some knowledge in simples, the possession of a musical clock, an electrical machine, and a stuffed alligator, added to a supposed connection with Thomas the Rhymer, lived for many years in very good credit as a wizard." One may well wonder what True Thomas would have thought of the electrical machine, not to speak of the stuffed alligator.

---

* Lest the reader should be unduly alarmed, I may say that the Clyde Trustees blasted this stone out of the way a considerable time ago, though whether this was done to prevent the prophecy coming true or, more prosaically, to clear the channel, I cannot tell.

† This part of the prophecy suggests that so many males are to perish in the fighting that polygamy will be a result, and women will be glad to take even a share in any available man.

A discussion of the esoteric fauna of the North would demand more space than is available here, so that only one or two examples can be touched upon. The Glaistig was a tutelary being in the shape of a little grey woman with yellow hair reaching to her heels, and dressed in green, the fairy colour. She was supposed to have been a mortal on whom fairy nature had been conferred. Solitary by habit and seldom seen, she haunted castles and houses. Generally, she was helpful or harmless, though occasionally obnoxious ones were found. Her useful functions were the overseeing of the cattle and of the way in which the house-servants performed their tasks.

Allied to the Glaistig is the Gruagach, but the latter is found as both male and female, the former sex being devoted to cattle and sheep folds.

Then there is the Brownie, which provides one of the puzzles of this intricate subject. The belief in the Brownie seems to have been Teutonic and not Celtic, according to Gregorson Campbell* who collected his information purely from oral accounts and was probably nearer the true versions than authors who depended on printed sources to some extent. Thus, the Brownie is not known in the Highlands, but seems confined to the tales current in the Lowlands or Orkney and Shetland where the settlers belong to the Teutonic race. In Argyllshire, stories of the Brownie may be due to quite recent infiltrations of southern influences.

A picture of the Brownie is given in some verses by William Nicholson, who was born in Galloway in the eighteenth century; they tell of the arrival of the Brownie called Aiken-drum at a farm house:

> His matted head on his breast did rest
> A lang blue beard wan'ered down like a vest;
> But the glare o' his e'e hath nae bard exprest,
>   Nor the skimes o' Aiken-drum.

---

* John Gregorson Campbell, *Superstitions of the Highlands and Islands of Scotland* (1900).

> On his wauchie arms three claws did meet,
> As they trail'd on the grun' by his taeless feet;
> E'en the auld guidman himsel' did sweat,
> > To look at Aiken-drum.
> "I lived in a lan' where we saw nae sky,
> I dwalt in a spot where a burn runs na by;
> But I'se dwall noo wi' you if ye like to try—
> > Hae ye wark for Aiken-drum?"

In another set of verses he is described: "There was a man cam' frae the Moon, And his name was Aiken-drum."

Glaistig, Gruagach, and Brownie were all supplied by the householder with milk which was sometimes poured out into special hollows in rocks.

As Gregorson Campbell pointed out, an origin of the superstition is not easy to reach. Obviously, a belief in such house-spirits can have arisen only after the tribes ceased to lead a wandering existence and settled down into some fixity. Probably it added to the prestige of a house if a supernatural being were supposed to be attached to it; and the owners may not have been adverse to the idea of an unseen creature which helped to keep servants up to their work; whilst lazy servants may have been ready to adopt the screen of a mischievous supernatural in order to account for their own misdeeds.

J. F. Campbell, in his *Tales of the West Highlands*, suggests that the Gruagach might be a Druid fallen from his high estate and living on the milk left for him by those whose priest he had once been, or, alternatively, that he was a half-tamed savage with long hair and skin clothing, hanging about the dwelling for what he could get. Gregorson Campbell, however, will have none of this, pointing out that the traditional appearance, character, dress, and actions of these beings are incongruous with the idea of Druid, heathen deity, or savage.

One point is interesting. These creatures resemble in nature the cat rather than the dog. The dog attaches himself to a person and not to a place; whereas the cat's affinity is for the place, irrespective of the temporary owner. In the same way, it is the house which attracts the Brownie, and even if the quondam owner is replaced by someone else, the Brownie remains in his old quarters.

Unlike the Brownie, the Urisk haunts unfrequented and mountainous spots and seldom comes near mankind. In the summer it is supposed to live in the high remote corries and to come down into the straths in the wintertime. It was usually seen in the evening, big and grey, sitting on the top of a rock and peering at intruders on its solitude, or slowly moving out of the stranger's way. Sometimes it gave safe convoy to belated walkers. It seems to have frequented the neighbourhood of sequestered waterfalls and mountain streams, but it had nothing in common with the southern Kelpie, which is unknown to the Highlands. Tradition has it that the Urisk, male or female, is the offspring of a commerce between a mortal and a fairy.

The Urisk seems to have persisted into quite recent times. Gregorson Campbell, who died in 1891, writes that "not many years ago" what was believed to be a Urisk was seen at sunset in Strathfillan, in Perthshire. That year, some sheep unaccountably disappeared from the hill and a quantity of grain vanished from the barn of the neighbouring farm.

It is curious to note that the currency of tales about the kindly or neutral supernatural has persisted in the Highlands, whilst legends depicting malevolent beings have declined. The existence of the loch-haunting Waterhorse, for instance, is falling into discredit. It was supposed to be a creature capable of appearing in the form of a horse or in partly-human guise with hooves. Its lacustrine origin was betrayed by fragments of green weeds and sand which clung to it. When it picked a man for its victim, it assumed equine form, induced the man to mount it, and then rushed into the loch; when it cast its eye on a girl it took the shape of a man, seduced the chosen maiden and then carried her off into the waters. In both cases, the victim was devoured and fragments of the body were washed ashore.

In one respect, Celtic folklore comes into line with other branches of the subject, since in it also the fairies are depicted as living underground in hills, knowes, and hummocks carpeted with vividly green turf. (*Sithein*—pronounced Shee-en—is the generic name for any fairy dwelling-place.) When True Thomas with his fairy mistress took "yon bonnie road that winds about the fernie brae," they rode into a "desert wide"—

> "And they saw neither sun nor moon,
>   But they heard the roaring of the sea.
> It was mirk, mirk night, there was nae starlight,
>   They waded thro' red blude to the knee;
> For a' the blude that's shed on the earth
>   Rins through the springs o' that countrie."

So avers the traditional ballad *Thomas the Rhymer* and the first three of the foregoing lines certainly suggest the subterrene. The last three, it may be pointed out, are by no means out of harmony with the bloody-minded cast of Thomas's own prophecies. And the troglodyte existence receives further support from the legends of the Rhymer still dwelling inside Dumbuck Hill or Tomnahurich.

Various suggestions have been made to account for these fairy legends on the assumption that the fairies were really survivals of an earlier race, such as the Picts, lurking in rock clefts among the high corries and keeping themselves under cover when any stranger approaches their dens.* I have heard this idea described as "impossible," on the ground that nothing of the sort could have escaped the attention of the Ordnance Survey map-makers. Perhaps the critics might change their views if they would take the trouble to go into the lonelier parts of the Highlands, or even into that desolate tract of country round the Merrick, north of Loch Trool in Kirkcudbrightshire. After the rising of 1745, the Highlands were

---

* If the reader has a leaning towards the gruesome in fiction, he will find examples of this theme in *The Three Impostors*, by my friend, Mr. Arthur Machen, and in *The Watcher by the Threshold* by John Buchan.

pretty thoroughly combed by the Army, but it is surprising to find how much escaped that net.

For my own part, though I have no belief in the survival into the present day of scattered Picts living their old life, and perpetuating their ancient customs among remote and savage hills, it is futile to deny that the thing is possible, if not probable. Much more likely is it, however, that something of the kind did actually happen long ago and that some record of this has been retained, vivid as of yesterday, engraven in the tenacious structure of Celtic memory. The modern city dweller, submerged in a rush of trivial happenings, can hardly realise how durable memories may be when existence is simpler and people have more mental leisure.

Curiously enough, as I was writing the foregoing paragraphs, I came across an example of this. One of my Highland friends comes of a family belonging to Glen Lyon, and from her I learned the following anecdotes which have been handed down from father to son among her kinsfolk. It will be best to give the details first and reserve till later any speculation as to the actual date of the occurrences.

There is no fox-hunting etiquette in Scotland, where foxes are treated as vermin and are exterminated by the gun or the trap. The nature of the countryside furnishes quite sufficient justification for such "unsportsmanlike" methods. In the old days, the eradication of foxes was to some extent handed over to professional fox hunters who knew every inch of the terrain over which they pursued their calling. One of these was employed in the Rannoch Moor district which abuts on the head of Glen Lyon, and he had his dogs to help him in his work. When the motorist on the Glen Coe road has ascended the Black Mount and gazes out across Rannoch Moor towards Loch Laidon he sees a landscape like something abandoned in the Creation as too unprofitable for any use: a dreary, tangled waste of bog, rock, pools and heather stretching east and northward towards far-distant hills. Even to-day it looks uninhabited and uninhabitable in the old times it must have been still more of a desert. This was the kind of country which the old fox hunter had to cross and recross in his journeys from one hunting ground to another.

The story goes that a friend once asked him if he had never been frightened in his rovings through this desolate region; and he confessed that thrice he had come upon things which made his blood run cold.

The first happened thus. In those days, here and there on the bleak face of Rannoch Moor a few huts had been erected to give shelter to the benighted wayfarer, and to one of these the fox hunter had made his way. As he came near it, he had the feeling that something was wrong; but he pushed open the door and kindled a light. As he did so, he found himself confronted by the body of a "redcoat"—a soldier—who had hanged himself with a rope from the roof beam. That was certainly no comfortable sight for any man to come upon, far out on the lonely Moor and in the darkness of the bothy. No great wonder that it chilled the spine of the fox hunter.

The second instance was, if anything, more gruesome still. While making his way up a glen in the dusk, he was struck by the behaviour of his dogs, which seemed perturbed by something which he could not perceive. Then, in his tracks, he heard a creature running with soft feet, pad, pad, pad, pad; and the dogs gathered about him with bristling coats. The sound of running feet drew closer and then, suddenly, he saw the figure of a naked woman, with flying hair, dash past him and vanish into the gloom ahead, leaving him aghast at this uncanny apparition. When he had got over his momentary panic, he set out for the nearest human habitation, told his tale, and got search parties organised. A day or two later, in the depths of a wood, they came upon their quarry: the body of a poor mad woman who had broken away from her friends and, pursued by the demons of her imagination, had run and run, into the wilderness until she fell exhausted and her life went out.

In these two cases, the fox hunter had come upon things which, though nerve-shaking, were still explicable; but his third crisis took a different shape. One day, while far away from the nearest aid, his dogs betrayed signs of uneasiness. Then, without warning, these familiar beasts turned upon him and did their best to pull him down. Any of my readers who are dog lovers can imagine his horror and consternation when, with no visible reason, his friendly beasts were suddenly transformed into a pack of dangerous enemies and he was plunged into a fight for life with creatures who, normally, were his surest guards. By surprising good fortune, he managed to beat them off at last, and eventually they grew normal and obedient as before. But while the frenzy lasted, it must have been appalling in its danger to an isolated man; and the apparent causelessness of the transformation must have added more than a tinge of the macabre to the grimness of the physical hazard in which he stood.

Now turn to the probable date of these stories. My friend tells me that in her family they can be traced back well over a hundred years; but I have a suspicion, based on the internal evidence of the tales themselves, that they go a good deal further back than a century. The reader can form his own conclusions from the evidence. Two points seem significant: the suicide of the "red-coat" and the obvious fact that the fox hunter carried no gun during his battle with his dogs. How came a soldier into the wastes of Rannoch Moor, far from any place where a garrison existed in recent times? One might postulate that the man was a deserter, but if such he were, why should he have come so far and why did he plunge into that wilderness in order to commit suicide?

A more helpful line is opened up if we inquire when "redcoats" are actually known to have been employed upon Rannoch Moor; and the answer to that takes us back to the days not long after the collapse of the Young Pretender's campaign at Culloden in 1746, for then the whole countryside in that district was dotted with pickets and patrols of troops incessantly combing the hills, moors, and glens for fugitives from the defeated army of the Prince. At that period—and only at that period—is the presence of a red-coat soldier easily explicable, and it will be noticed that in the tale the fox hunter was staggered merely by the hanging corpse and not by the fact that it was the body of a soldier, many of whom he would have met in the district at that period.

What brought the wretched "red-coat" to suicide? Probably he had got into trouble and stole away to put an end to himself, a thing easy enough to achieve during those particular operations.

Perhaps, however, there is another explanation. Cumberland's troops were drawn largely from the South and many of them had never been in mountainous country before. Now it is a fact that though many people love the hills, there are marked exceptions who feel a curious uneasiness if they are placed in lonely tracts. It is not a matter of agoraphobia in its ordinary form; it is a peculiar disquietude, arising perhaps, from a wholly unfamiliar environment. Being myself a lover of hills, I can give no better account of a sensation which I have never experienced personally. Suppose, however, that the unfortunate "red-coat" suffered from this disquiet in addition to the ever-present threat of an attack upon an isolated soldier by some desperate Jacobite band. Might not his trouble have grown upon him until he could bear it no longer and took the short way out? If not a probability, this is at least a possible ground for the suicide, away in that remote and desolate tract.

Then take the second point in the tales: the fact that the fox hunter carried no gun. After the '45, all Highlanders were disarmed, and none of them could carry a firearm openly. So this also would tend to suggest the same period.

The reader can make up his own mind on the matter. The undoubted fact is that the tales came down by oral transmission for over a century in the same family, and that there is no proof that they had not already been handed on from father to son before that, as far back as the post-Culloden period. In any case, they give a good instance of the retentiveness of the Highland memory.

These are far-off things. Now comes something much more modern, but first some account must be given of the man who told me the story.

In my time, I have met many eminent people; but in looking back over a long career I can think of none, who, as a man, impressed me more than my old Chief, John Norman Collie,* with

---

* A fuller account of Collie will be found in the *Obituary Notices of Fellows of the Royal Society*, Volume 4, November, 1943, pp. 329-356.

whom I was associated—greatly to my advantage—during my early years in chemical research, and whose friendship lasted till his death, forty years later, in 1942. He taught me much in my own particular branch of science, for he had the big simple ideas which are the mark of a fine mind, and his manipulative skill was a joy to watch; but chemistry was the least of the things which I learned from my association with him.

Collie was of Highland descent on one side of his family, and he had some of the characteristics of the race. Mr. St. John Ervine wrote of him: "Collie was not communicative until he knew you. Then his conversation was rich and ample. . . . A first impression of Collie would have been of a very silent solitary, aloof and almost unfriendly man; but he could be very good company; and I was struck, every time I saw him, by the liking he inspired in young people." Certainly he was adored by his students; and if I had been in a physical or moral "tight corner," Collie was the man of all others whom I would have chosen as a stand-by.

He received more than the usual share of academic honours, one of them being conferred on the recommendation of Dr. Nansen. At the age of thirty-seven he was elected a Fellow of the Royal Society. He was a Vice-President of the Royal Geographical Society and succeeded Sir Francis Younghusband as Chairman of the Mount Everest Committee; and he was President of the Alpine Club.

He knew a good book and a good wine; he played a creditable game of golf; he was a first-rate shot; and, as a fisherman, he roused the exasperated envy of fellow-anglers by practically clearing burns in which they could hardly raise a nibble.

But his real interest was in beauty, whether it was found in a crystal or on a Himalayan peak. Once, during Collie's investigations of the behaviour of gases under the electric discharge at low temperatures—a by-product of which was the neon light which diversified our streets in non-blackout days—Sir Herbert Jackson came into his laboratory and, after watching for a time, said: "Collie, I truly believe that you are far more interested in the colours of the discharge than in the striking phenomena you are observing." And that was probably true.

His collection of specimens of Chinese and Japanese art made his rooms a museum; and I remember hours spent there after dinner when, being led on to talk of porcelain or carving or lacquer, he would illustrate his theme by picking up an appropriate piece to point out its excellences. When, towards the end of his life, he weeded his collection of its less valuable items, even these rejected articles fetched thousands. If the talk turned on literature, it was the same thing; he had something at hand to drive home his point.

He was a good painter; and as a landscape photographer he was in the front rank. Some examples of his work in this field are to be found in his books.

His great-uncle was the discoverer of the Collie River in Western Australia, and his statue was unveiled in the town of Collie, named after him, at its centenary in 1935. Thus, Collie had exploration in his blood: and it came out. He was one of the small group of pioneers who initiated guideless climbing in the Alps. In 1895, he joined in the first attempt to climb Nanga Parbat in the Himalaya, during which Mummery lost his life. It was no picnic, that experience on the greatest sheer mountain face in the world. Collie, in London, had a fine taste in food and drink; but he delighted to put all that behind him at the call of the hills and to go out and live amid the roughest conditions. I remember one evening in his rooms, after a dinner fit for a Sybarite, watching the appreciative twinkle in his brown eyes as he described how four of them ate a whole sheep, after two weary days and nights spent starving on the icy and inhospitable slopes of Nanga Parbat. "But the sheep there are rather small," he urged in extenuation of this Gargantuan meal.

After the Himalaya, he turned to the Canadian Rocky Mountains, to which he went back year after year until he had mapped by plane-table survey hundreds of square miles of hitherto unknown country.

Not only could he do these things, but he could talk and write interestingly about them, salting his descriptions with his own acid humour. He had an excellent style and a gift for the apt phrase.* It is

---

* His *Climbing on the Himalaya and Other Mountain Ranges* shows him at his best, his other volume *Climbs and*

no great wonder that a distinguished friend of his once remarked: "Collie, it seems to me that you are a chemist only in your spare time."

He has left his name graven among the hills. There is a Mount Collie in Canada; Collie's Step in the Moss Ghyll climb is known to all who have gone that way; and in Skye there are several routes known as Collie's climbs, whilst one of the peaks has him as godfather: Sgurr Thormaid or Norman's Peak.

It was to Skye that he ever returned and it was thither that he went at last to die. In Hillary's *Last Enemy* there is an acute picture of Collie at Sligachan, "the old mountaineer," unnamed and described by one who never even spoke to him, but who was none the less impressed by his personality. Author and subject are both gone now. Collie lies in one of his familiar glens; a little burn tinkles by his grave; and around him are his well-loved hills.

Now I come to Collie's experience with the Fear Mohr on Ben Macdhui in the Cairngorm Mountains. I give it as he himself told it to me, with two supporting pieces of evidence. A condensed version of the narrative is printed in the *Royal Society Obituary Notice* of him, written by Professor E. C. C. Baly.

One day during an Easter vacation, Collie set out alone to climb Ben Macdhui. There was snow on the ground, and it was misty weather. He had a rucksack strapped on his back. Not far from the summit, he came upon a clear stretch of snow, over which he continued his walk, when suddenly, behind him, he heard footsteps. He thought little of that, merely supposing that some other climber was on the hill. Nothing was visible on account of the mist wreaths. After a little while, Collie paused, expecting the other man to make

> (cont.) *Explorations in the Canadian Rockies* (written in conjunction with H. E. M. Stutheld) is, by comparison, rather in the nature of hack-work written merely because some account of these explorations had to be presented. Other literary work of his is to be found in *The Alpine Journal, The Scottish Mountaineering Club Journal, The Geographical Journal*, and *The Cairngorm Club Journal*.

up on him; but as soon as he halted, the noise of those other steps stopped also, and there was dead silence. After a moment or two, Collie moved on again, paying no particular attention to the matter. Then he heard the footsteps again—an unmistakable crunch . . . crunch . . . crunch in the snow-crust.

Rather puzzled, he turned back in the direction from which he heard the sound, but he found no one; and again he continued his way up towards the summit. Up to this moment, nothing had occurred which seemed in the least abnormal; but at this stage a peculiarity in the dogging footsteps forced itself on his attention. They were keeping level with him, yet the walker was taking but one step to Collie's two, and only a giant could have done this.

At this point, Collie felt a terrible fear flood over him. "It was just panic. I never felt anything like it, before or since. I didn't look back. I simply turned and ran down the hill, and I don't think I stopped running until I came near the outskirts of Rothiemurchus Forest. Nothing on earth will take me up Ben Macdhui again, after that experience. I can't explain it."

But he added to his own tale an episode derived from the experience of another climber whom I knew slightly. This was Dr. Kellas, who was leader of one of the Mount Everest Expeditions and who, unfortunately, died during that work.

Kellas, it appears, was one summer camping on Ben Macdhui with a companion. They were sitting at the edge of the plateau, watching the sunset: Kellas beside their camp, his companion further along and not far from the cairn. It was a beautiful evening. As he sat there, Kellas noticed a man's figure making its way diagonally up the slopes below the crest; but he was intent on the sunset and paid no special attention to the approaching figure, merely recognising that it was there, without examining it, as most of us would do in the same circumstances. The figure came on, breasted the edge of the plateau, turned away from Kellas towards his companion, and walked on. It circled round the cairn and returned, passing between the cairn and Kellas's companion and quite close to the latter. Then, at about the same point as it had mounted the plateau, it turned off and began to descend the slope below, up which it had come.

I may point out that Kellas felt no unwonted sensations during this transit. Still intent on the sunset, he had perceived the figure and subconsciously noted its doings; but it had not attracted his direct attention in the slightest; and I believe that up to that point he had taken it for some shepherd with business on the hill. But when the sun went down, his attention turned from the landscape and he recalled the figure again. He remembered then something which he had seen, without paying any attention at the moment: when the figure came level with the cairn, the two objects were about the same height.

Now the cairn is about twelve feet high; and when he recalled these facts, Kellas's attention was aroused. He walked along and, joining his companion, said: "That was a tall fellow who passed you a little while ago." His companion was surprised. Nobody passed me." Kellas then looked down the slope to find the figure and call his companion's attention to it; but by that time it had disappeared.

A third piece of evidence bearing on the same subject comes from Mr. Colin Phillip, the artist, one of whose hobbies is to pick up information about local superstitions from the Highlanders. In chatting with an old native of the Cairngorm, Mr. Phillip turned the talk to Ben Macdhui and inquired if there was anything of interest on the hill. The answer was: "Och, aye, there wass aalways a Lang Man on Ben Macdhui."

There is the tale as it came to me. I have met all the three witnesses, and I have not the slightest doubt as to their complete veracity. Short of seeing the thing with my own eyes, I could ask for no better evidence. All three were cool, unemotional people, with a far better knowledge of the hills than falls to the lot of the average man; and none of them had any previous knowledge of the legend of a giant on Ben Macdhui.

A note or two on the matter may not be amiss.

In the first place, take the fact that when Collie halted, the sound of "the other footsteps" ceased at once. Now leather straps in equipment sometimes creak when the wearer moves, but the noise ceases when he stops walking; and it might be suggested that sound came from the straps of Collie's rucksack. Very little consideration is enough to dispose of this hypothesis. The creaking of a leather strap is quite different from the sound of feet crunching in snow; and, after the account I have purposely given of Collie's long experience in ice and snow among the high hills, no reasonable reader will be inclined to believe that he would be in the least likely to confuse two such sounds with each other. Further, if a creaking strap were at the root of the affair, Collie would have noticed it long before he came near the Ben Macdhui summit.

Again, extreme exhaustion sometimes produces queer effects. A friend of mine was in the retreat from Mons. He saw no angels, though he did see a horse and its rider in front of him simultaneously fall asleep and come down with a crash on the road; but he said that when men reached that stage of weariness, he would not be surprised by them "seeing" anything, from angels downwards. We know, too, from Herodotus, that when Pheidippides ran from Athens to Sparta (roughly 150 miles) in less than two days, he encountered the God Pan *en route*, which is not at all surprising to hear, in view of the man's probable exhaustion. But in Collie's case, exhaustion does not enter the problem; no man of his fitness would be fatigued by the exertion of climbing Ben Macdhui.

It seems to have been a case closely corresponding to the old Greek experience of Panic fear; but to say that is no explanation but a mere restatement of the thing in other words, since the cause of the Panic fear is still a mystery. It is certainly curious that the solitary walker should be attacked by panic, though he saw nothing; whereas Kellas, in company, felt no uneasiness, even though he actually saw something out of the common.

Perhaps some of my readers may be inclined to repeat Collie's expedition at the same season and under the same conditions, in the hope of throwing further light upon the matter.

This essay may be fittingly closed with an account of my own experience of that curious phenomenon conveniently described as "The Loch Ness Monster." Being well aware of the danger of diverging from the truth after telling a story several times, I would hesitate to describe the affair from memory after the lapse of a decade. But, as it happens, on my return home after seeing the thing—whatever it was—I sat down and wrote a short account of it to Lieutenant-Commander R. T. Gould, R.N., who had published an exhaustive book* on the Monster a year before; and in stating the facts here, I shall confine myself to what I set down in my correspondence with Commander Gould. Thus, one source of error will be removed from my tale.

On September 4, 1933, I was motoring up from Fort William to Inverness along with my wife and daughter. On the earlier part of the road, we had rain and mist, but by the time we reached Loch Ness the weather had improved and every detail on the opposite side of the Loch was clear and sharp.

About 6 p.m. we were a mile or two short of the place where the road bends inland from the Loch at Drumnadrochit, and ahead of us was another car containing a man, a woman, and a child—strangers to us. I may say that we were not thinking of the Monster, nor did it come into our minds when the car ahead of us stopped suddenly and its occupants got out hurriedly and rushed across the road, evidently excited by something. Out of mere idle curiosity, we pulled up our car some yards ahead of the other, and looked out over the Loch to see what had attracted the attention of the other party.

Several hundred yards out in the Loch, we saw quite distinctly a large patch of foam and, in it, some black things moving. (I am a poor judge of distance over water, so do not care to give a figure which would probably be erroneous.) My wife exclaimed, rather contemptuously: "It's only a school of porpoises"; which described the appearance exactly. Speaking for myself, I saw four (or five) "coils" or "humps" in motion among the foam. These remained in sight for perhaps a minute, and then vanished so far as I was concerned. The foam patch remained, however. It was not a

---

* *The Loch Ness Monster and Others*, by R. T. Gould (1954).

"white-cap," for there was very little wind at the moment, and the foam remained in one spot and did not drift as a "white-cap" does. It resembled the breaking of water over an almost submerged reef more than anything else that I can think of. We watched for a few minutes but saw no reappearance of the black coils or humps.

I should like to be definite on the following points: (1) The air was perfectly clear at the time, after the rain; there was no obscuring mist; and the light was good for 6 p.m. Summer Time. (2) We had not the Monster in our minds until we actually saw the thing, so we were not in any state of expectation which might have misled us. (3) What we saw left in our minds a decided impression of something large and active, and was not in the least like an old tree trunk or anything of that sort. (4) All three of us saw it and compared notes afterwards. (5) We did not discuss the matter with the three people in the other car, nor did I even speak to them, so that our observations were quite unbiased in this respect.

On the following Monday (September 9) an Inverness paper published an account which evidently described what we had seen. Sheriff Principal George Watt of Drumbuie, Glenurquhart, was motoring to Inverness in company with his wife and daughter. The Monster appeared, they said, "a mile east of Temple Pier, and was splashing a good deal." The Sheriff did not get a good view of the creature. Mrs. Watt said she saw four or five humps. Miss Watt thought there were six or seven.

Apparently, then, this particular avatar of the Monster was seen by three independent groups of people: at least eight witnesses in all; and mass-hallucination seems to be excluded.

The same issue of the Inverness newspaper chronicled a "second view," as follows: "Another appearance was reported from the east end of the loch near Dores by three people motoring towards Inverness. They are Mr. R. Stone, a chauffeur from London who is employed by the tenant of Killin shootings, Stratherrick; a cook, Mrs. Bird, also from London, and a friend. They stated that they saw a large dark body emerge near the Dores side and travel up the loch with a lurching motion."

These, then, are the facts anent my solitary experience of the so-called Loch Ness Monster. It will be noticed that one of my companions likened the appearance to that of a school of porpoises; and the question arises: could porpoises make their way into an inland sheet of water like Loch Ness and could they live in the fresh water when they reached it?

Commander R. T. Gould, after a careful examination of the various possible modes of ingress from the sea into Loch Ness, stated in his book:* "It appears at least quite possible that, during a heavy spate, a creature drawing not more, say, than five feet of water (and, necessarily, a powerful swimmer) might make its way unobserved, by night, into Loch Ness." And porpoises can, apparently, exist in rivers at points so far distant from the sea that the water can hardly be regarded as salt. They have been seen, for instance, in the Thames at Richmond, and in the Seine as far up as Neuilly. But even assuming that a school of porpoises made their way into Loch Ness and excited attention by their gambolings, it seems doubtful if they could present all the various phenomena which demand explanation. Commander Gould, certainly, will have none of this hypothesis.†

For my own part, I must content myself with saying that my experience of Highland lochs is fairly extensive; but in none of them, before or since, have I seen anything akin to this object which I saw in Loch Ness on the occasion which I have described.

If I were pressed further, I should take refuge in repeating a remark made by my sagacious friend, Mr. Arthur Machen,‡ when he was consulted upon a refined point in stage craft by that famous actor-manager, Sir Frank Benson. "Well, Mr. Machen, what do you think about it yourself?" "Indeed, sir, I don't venture to have any opinion." If anyone asks me what I think about the nature of the Loch Ness Monster—and some other things as well—I copy Mr. Machen. "Indeed, sir, I don't venture to have any opinion."

---

* *Op. cit.*, p. 10.
† *Op. cit.*, p. 117.
‡ *Things Near and Far*, Caerleon Edition, p. 166 (1923).

AFTER DEATH THE DOCTOR

AFTER DEATH THE DOCTOR

"Better go in, now, doctor?" Sergeant Longridge suggested. "You'll be here later on?" he added to Jack Sparkford. "I'll have to ask you a few questions then."

It was the sergeant's first murder case, and though outwardly confident, he felt a shade diffident in the presence of the more experienced police surgeon.

Hastily summoned, Longridge had tramped up to the house, breakfastless, planning his procedure as he came. The first thing to do, obviously, was to examine the room and the body, with no outsiders to worry him while he made his notes. It was with some relief that he saw Dr. Shefford's car drive up to the front door as he reached it himself.

Here was the room, with that comfortless and untidy look which rooms have in the morning, before the disorder of the night has been repaired; the chairs set at odd angles, the ashtrays unemptied, the cushions crumpled and awry, a newspaper thrown carelessly on the floor.

It seemed a cross between a sitting-room and an office, with a wireless set, bookcases, filing cabinets, a couple of occasional tables, and a pedestal desk against the wall in one corner. The furniture was good, but shabby like the clothes of a man who has come down in the world.

"Regular old bachelor's den," the doctor observed. "Not a flower anywhere, though they've plenty in the garden."

He put down his bag and went over to the corner of the room. The sergeant was not sensitive, and it cost him no qualms to examine the body of Barnaby Leadburn as it lay back in the office chair before the pedestal desk. A bit weird, he reflected, to see an old gent one knew by sight, lying there with his throat cut.

Dr. Shefford's interest was more professional.

"Not so much blood as I'd have expected," he commented.

His examination of the body became more technical; and Longridge, understanding little of what the doctor was doing, bethought himself that he had a task of his own. He opened his notebook and began to jot down things which seemed important.

The French window was open, but only one of the curtains was drawn back from it. The electric light had been switched off. In the empty grate were some ashes of burnt paper, with printing showing black on the grey background: a piece of newspaper, the sergeant guessed. And when he picked up the newspaper from the floor, he found about a quarter of one of the sheets torn away.

The desk bore a neat array of account-books, and on the open pages of the one nearest the body lay a sheet of note-paper. Longridge, craning over the doctor, read on it the words: "*I, Burnaby Leadburn . . .*" Then came the regular trail of blotches caused by a pen rolling along the paper; and the pen itself lay on the desk surface to the right of the book.

For some moments Sergeant Longridge puzzled over this sinister hiatus in the manuscript, trying to imagine how the incomplete sentence had been meant to run. At last an idea occurred to him.

"Do you think, sir, that he could have been making his will? I never made a will myself—not worth while, seeing I've nothing to leave—but don't they run: I, So-and-so, hereby give and bequeath . . .' That's how I've heard say."

Dr. Shefford seemed to think that he had enough to do in his own sphere, without trenching on the sergeant's.

"They usually start by appointing executors," he said dryly. "This looks a very clean cut, sergeant. It might almost have been done with a razor, from the look of it."

"How long do you think he's been dead, sir?" Longridge inquired, glancing at his watch.

Dr. Shefford shrugged his shoulders rather impatiently.

"He probably died round about midnight, I should guess; but it's no use pretending that you can tell to a minute from medical evidence alone. Things vary far too much for that. Hadn't you better hunt about and see if you can find the weapon it was done with, if it happens to be in the room?"

The sergeant's feelings were ruffled by this reflection on his zeal and efficiency.

"Well, sir, if you'll stand aside for a moment, I'll have a look with my flashlamp in the well of the desk. I don't see any weapon lying about anywhere else."

Dr. Shefford stood aside; and the sergeant, cautiously groveling with his flash-lamp, explored the cavity under the desk. An exclamation of triumph told the doctor that something had been found.

"Here's what it was done with, sir. Look! That's a rummy sort of knife. I'll fetch it out."

Gingerly he picked up the weapon and placed it on the desk: a stout blade four or five inches long embedded in a straight wooden handle, with a steel lever at the side.

"It's the sort of knife artists use for cutting mill-board or for trimming prints," the doctor explained. "You can alter the amount of blade that sticks out of the handle by setting that steel catch on the side. I rather wondered how that wound was made, but this evidently did it, to judge by the blood on the blade. By the way, sergeant, you'd better inquire if the old man was left-handed. This gash has been made from right to left, by the look of it. If old Leadburn was right-handed, then it isn't suicide. And you might ask if the light was on or off this morning. Suicides aren't usually so economical as to switch off before they put their own light out. Though, from what I've heard, old Leadburn was mean enough to have taken the precaution."

"Very good, sir," the sergeant agreed. "And now, sir, I think I'll leave you and make some inquiries from the house people."

"Just pull the curtains before you go, and switch on the lights. It's hardly decent to carry on further in full view of anyone who happens to cross the lawn."

Jack Sparkford and his younger brother, Sydney, were in the hall as the sergeant emerged from the room. Between Jack, at twenty-five, and the fifteen-year-old schoolboy, the family likeness was unmistakably even down to the weakly obstinate chins.

"They look bothered, but not just tearful," Longridge reflected. "Not much wonder, either, if old Leadburn was the tartar he got the name of being. They hadn't the lives of dogs, if all tales were true."

"Have you found anything?" Jack demanded anxiously.

The sergeant put up his hand defensively.

"One thing at a time, sir. We've got no facts yet. I'd like to see the maid who discovered the . . ."

"Ring for Hart, Sydney," Jack ordered. "We'd better go in here. No use interviewing her in the hall."

In a few moments a young, rather good-looking housemaid appeared, evidently in a very shaken state of nerves.

"Your name's Jenny Hart, isn't it?" demanded the sergeant. "Tell me just how you came to discover Mr. Leadburn this morning."

The girl seemed taken aback by this official tone from a person with whom she already had a nodding acquaintance.

"It was this way," she explained nervously. "I went into that room about seven o'clock to clear it up and set things to rights."

"Was the electric light on?"

"No. So the first thing I did was to go and draw the curtains back from the window, and when I'd drawn the first ones I saw Mr. Leadburn lying there in his blood on the chair, and I screamed and ran out of the room."

"Was the French window open when you drew the curtain?"

"Yes, it was. It must have stood open all night. Mr. Leadburn always liked to have it open, but he used to shut it when he went up to bed, last thing."

The sergeant could think of no further questions to ask the maid just then, so he dismissed her and turned to the two nephews of Barnaby Leadburn.

"Did anything out of the common—sounds or what not—attract your attention in the night?" he asked Jack Sparkford.

Jack shook his head.

"I went up to my room about eleven o'clock," he explained. "My uncle was busy with his accounts and so forth. I heard nothing suspicious."

"And you, sir?" Longridge inquired, turning to Sydney.

"I heard nothing inside the house," the boy answered at once. "But I heard Caesar—that's our big retriever, you know—I heard him give a long howl, a funny sort of noise, about twenty-past twelve. I never heard him make a noise like that before—a kind of howl, very long-drawn-out."

"How do you know it was twenty-past twelve then?" demanded the sergeant sceptically.

"Because the down express passed almost immediately afterwards. It goes through at 12.25. I was kept awake part of the night with toothache, you see. And, by the way," he added, "I haven't seen Caesar this morning."

"You'd better get hold of him," Jack said at once, "he might do damage to somebody if he's left on the loose."

"Savage, is he?" inquired the sergeant, who had heard some rumours about the dog.

"A bit nasty with everybody barring ourselves, I'm afraid," Jack admitted. "My brother Timothy's the only, outsider he takes to. Even the gardener at the lodge is in terror of him. Caesar hates him for some reason or other. My uncle had him let loose in the grounds at night. Nobody would dare to come a-burgling here with Caesar off the chain. It would be as much as his life was worth."

"What the girl said about open windows is right?" inquired Longridge.

"Oh, yes. We've all got a tendency to consumption in our family. My brother Timothy's in a bad way with it. Naturally we believe in fresh air, and my uncle was all for open windows."

"H'm!" said Longridge. "Now about this dog, sir. Had your uncle any enemies, that he kept a savage dog about the place at night?"

"Not that I know of," Jack replied frankly enough.

Something in Sydney's expression caught the sergeant's eye, and he put the same question to the boy.

"I don't know about enemies," Sydney answered doubtfully. "He had a bit of a row with Corfe—that's our gardener—last night. Something to do with our housemaid, Hart. I heard the two of them slanging each other and Corfe seemed a bit above himself with rage. He's engaged to Hart, you know. Something about my uncle accusing her of stealing. I couldn't help hearing some of it, but I didn't listen on purpose. It was in the garden."

Sergeant Longridge veiled his interest in this new piece of evidence by changing the subject. He described the knife and asked if it belonged to anyone in the house.

"Oh, yes, it's mine," Sydney admitted promptly. "It always lies on the desk, in there. I use it for trimming the white edges off my Kodak prints."

"Oh, it's your knife and it always lies on the desk," Sergeant Longridge repeated mechanically, as he noted the facts in his book. "Thanks. Now another question, Mr. Sparkford. Did any visitors call on your uncle last night?"

"Visitors? Not that I know of," Jack answered. "Unless you call my brother Timothy a visitor. He came in after dinner and left again about ten o'clock. I saw him down to the road then. His eyesight's bad, and I went with him down the short cut to save him stumbling about in the dark."

"He left at ten p.m.," noted the sergeant. "Have you notified him about the death, sir?"

"Yes, I rang him up. But he was sick, last night he had to call a doctor in the small hours, he told me. Probably he's a bit shaky this morning and hasn't felt up to coming over here yet."

"He lives across the railway, I think?"

Longridge's question was merely formal. He knew Timothy Sparkford by sight and reputation; but he believed in getting "the evidence of a witness" to put down in his notes.

"Yes, he lives in Moss Cottage," Jack replied.

"I remember, sir. He moved over there after he had a bit of luck in the Sweep, didn't he? By the way, your uncle didn't seem nervous about anything when you saw him last night?"

Jack shook his head decidedly.

"Nothing of that sort, not a sign of it."

"Nothing missing, that you've noticed, sir? No valuables gone, or a safe opened?"

"Nothing whatever in that line, so far as I know. There's no safe in the house. He kept everything in the bank."

"I see, sir. Now I think I'd like to have your maid back again just to ask her a question or two more."

But when Hart was summoned, they learned that she had left the house, apparently to talk to the gardener; and it was a few minutes before she came back. When she reappeared the sergeant interviewed her alone.

"Now, my girl," he began in his most paternal tone, "you don't want to be keeping anything back in a bad case like this. It wouldn't do. People would begin to think there was something wrong, you see, if you did that. So just tell me what this bit of trouble was that you had with old Mr. Leadburn."

At the question, the prim maidservant vanished and in her stead a virago appeared, furious, bitter, and yet uneasy.

"If anybody says I stole anything, they're taking away my character, and I'll have the law on them," she burst out breathlessly.

"The truth of it is, I was dusting his desk and I happened to pull open a drawer, and a note dropped out, and I picked it up, and I was standing there with it in my hand when he came in. He'd been out there in the garden, sneaking behind the bushes, watching me through the window.

"Spying on me! That's a nice occupation for a man with thousands a year, I don't think! I wouldn't demean myself so. And then he swore I'd stolen the note, and he'd been missing money for some time, for he kept a good check on it, which was like his mean, miserly ways. And I said I'd never touched a penny of any money in the house, and *he* said he was going to send for the police and see I was jailed for it, for he knew to a penny how much I'd taken—fancy!

"And I went and told Simmy—my boy, he is, and we're going to be married in three months. And Simmy was real angry, as who wouldn't be, and he went and told Mr. Leadburn straight just what he thought about it, and then Mr. Leadburn said he'd have no thieves on his premises and if Simmy married me he could look out for another job, for he wouldn't be kept on at the lodge.

"And Simmy was angry, of course, for with things as they are nowadays it's not likely he'd get another job, and so we wouldn't be able to get married after all. And they got to words about it, and you can't wonder at Simmy, and that's the plain truth. You can ask Simmy and see if it isn't."

"I see," said Longridge non-committally, when the torrent ceased. "Quite so."

He had a shrewd idea that most likely there had been faults on both sides in that affair; but he had no desire to start inquiring into petty larceny in the midst of a murder case. He noted that there had been bad feeling between the gardener and his employer.

"Now, tell me," he went on, "did you see anything that might throw light on this business?"

The girl seemed to realise that she might have done more harm than good by her outbreak, and when she spoke again, it was in a cooler tone.

"I happened to be going through the hall last night; about half-past nine or so, it was: and I heard them at it, hammer and tongs, in that room. It was about Mr. Jack's engagement and an extra allowance for him to get married on. Mr. Timothy was there, and the old man was as mad as a hornet at the idea and wouldn't hear of it. And he seemed to be threatening Mr. Timothy, too, for I heard something about 'discreditable doings,' and you know what Mr. Timothy is.

"They were fair shouting at each other. That old man seemed to have a fair down on people getting married: Simmy and me, first of all, and then Mr. Jack and his girl. And he didn't seem to like the other thing any better, neither, if you go by the way he was storming at Mr. Timothy. One would think he expected everyone to live like monks and nuns."

"A bad quarrel, it sounded like?"

"Oh, of course they were always quarrelling, if it comes to that," Jenny answered. "What else would you expect, with that old skinflint holding on to the cash and doling it out to grown men as if they were kids? It wasn't a happy household, as you might say."

"No, I suppose not," Longridge agreed. "Now, just another question. Was Mr. Leadburn left-handed?"

"Not he," Jenny replied. Then, after a pause, she added, "It's Mr. Timothy that's left-handed. Was that what you meant? What do you want to know for?"

The sergeant was saved from answering by the hasty entrance of Sydney Sparkford, evidently in a state of excitement.

"We've found Caesar!" he exclaimed breathlessly. "He's dead, poor old dog! Poisoned, by the look of him. You'd better come and see him for yourself. He's in the shrubbery close to my bedroom window. That's why his howl waked me up, I expect, coming from so near at hand."

Longridge followed the boy to the shrubbery, where they found Jack Sparkford before them, staring thoughtfully at the dead dog. The sergeant got the impression that its end gave him more regret than the loss of his uncle seemed to do.

"What do you make of this?" he demanded, as they approached.

Longridge went down on his knees and examined the dog which, even in death, looked a formidable brute. As he bent close, the merest whiff of a familiar odour came to his keen nostrils: the scent of bitter almonds.

"This'll be a job for the vet, sir," he said, tacitly admitting his own lack of expert knowledge. "I'll just have a look round, though."

He ferreted about for some minutes without success, but at last he discovered the thing for which he had been searching: a piece of raw liver. By holding it close to his nose he managed to detect the faint smell of bitter almonds from it also.

"That's how it was done, sir," he explained in triumph, holding out the meat to Jack. "Just you smell it, sir: same smell as the dog's mouth has."

Jack evidently disdained a personal verification; but Sydney eagerly sniffed first at the dog and then at the liver.

"I know what it is!" he declared jubilantly. "Cyanide, that's what it smells of. Just the same as the stuff in my butterfly killing-bottle."

The sight of the dead retriever had led Longridge to revise some of his ideas abruptly. "An outside job," was his new verdict. The dog had been friendly with the house people, and with Timothy Sparkford.

Any one of them could have got at old Leadburn without this huge brute interfering. But an outsider would need to dispose of it before he could penetrate to its master. Cyanide? Well, that should be an easy enough poison to trace. You had to sign for that at the druggist's before you got it.

He was stirred from his musing by the approach of a fresh figure: Timothy Sparkford, whom the sergeant knew by sight. He came forward with the hesitating walk of a man suffering from very defective vision, and as he reached the group he peered closely at each face before he greeted his brothers.

With his shambling gait, low stature, powerful physique, and deep-set eyes, he had something about him which contrasted sharply with the appearance of his younger brothers; and when he spoke, it was with a certain truculence.

"Who's this?" he demanded, after he had scrutinised the sergeant at close quarters.

"It's Sergeant Longridge, Tim," Jack explained. "I rang up the police before I got on to you on the phone."

Timothy acknowledged the introduction with a casual nod.

"Has the old geezer kicked the bucket?" he asked brutally. "Best news I've had in a month of Sundays. No use being mealy-mouthed about these things, is there? No flowers, by request. That's the right spirit in this case."

The ribald tone made Longridge prick up his ears, and it occurred to him suddenly that not one of the people he had questioned had shown the slightest regret for Barnaby Leadburn's death. They had not vented their spite like Timothy, but their restraint had been almost equally significant. They had not thought it worth while to pay even a tribute of hypocrisy to his memory.

"Sorry I couldn't get over sooner," Timothy went on. "When I got home last night I started some electro-plating, and in the middle of it I got the most frightful attack of gripes and sickness.

Must have been the remains of some shrimp paste I had for supper. Been left open too long, I suspect. Ptomaine poisoning, likely. Anyhow, after the bout I had, I could barely crawl to the phone and ring up Dr. Ackworth. He came along, not over-pleased at being dragged out at half-past twelve, I gathered. All he did, when he arrived, was to stand around and let nature take its course. Anyhow, it's over now, and I feel a bit better. Still a bit shaky though."

Then, completely ignoring the sergeant, he took his brother's arm.

"Come along and tell me all about it."

Longridge was going to call him back and question him when on the drive he saw the police surgeon beckoning to him. He hurried off, leaving Sydney to join his brothers.

"Oh, Sergeant," Dr. Shefford said when they met. "You can rub one notion off your slate. This affair wasn't suicide. He was strangled first of all, and then his throat was cut to hide the marks of the cord. At least, so one may suppose. That's why he made no noise when he died. And, naturally, with the heart stopped, he didn't bleed as much as one might have expected from the wound. There'll be an inquest, of course. There'll have to be some arrangements made for a p.m., I expect."

"Very good, sir."

Sergeant Longridge's investigations lasted longer than he had expected, but a couple of days later he was summoned—not for the first time—to give an account of his stewardship to Inspector Dronfield. The inspector was deep in a study of various documents, and he rubbed his eyes wearily as Longridge presented himself. He was a tall man who concealed a natural alertness behind an air of lassitude.

"Not clear yet?" he grumbled. "Suppose we take it step by step. Systematically, I mean. Must have been either an inside or an outside job. That's obvious. Insiders first. Not the cook?"

Sergeant Longridge shook his head. That suggestion was absurd.

"No. She's got a first-class character and she's only been in that place a couple of months. She hasn't had time to raise a grievance big enough to account for the job."

"The maid, Hart, then?

"I don't somehow see her strangling the old man and then cutting his throat," the sergeant declared. "She might, but it's not my idea of her."

The inspector made a non-committal noise.

"Hardly sounds like the schoolboy, either," he confessed, "though one never knows what some kids may get up to in these days. That leaves Jack Sparkford as a possible."

He picked up a document from the table.

"Your stuff about the family's all gossip. Still, it seems pretty sound. I've had it checked up. Old Leadburn did get round his widowed sister before she died, and he drafted her will for her. She trusted him, it seems.

"Here's a copy of the will. Got it from Somerset House. Leadburn was to draw the income from the estate—about four thousand pounds a year—until the youngest son reached twenty-one. Then the four of them were to divide the capital in equal shares. Meanwhile, Leadburn was to allow each of them annually a sum equal to his own personal expenditure for the year. I suppose she thought that meant a very comfortable income for each of them.

"What happened was that he turned out to be a miserly old skinflint who lived on about a pound a week himself, and he paid his nephews at the same rate. Shows how the best intentions may go wrong. And he wouldn't let them take up any trade or profession.

"Coming into a thousand pounds a year apiece, later on, they didn't need it, he said. Then there's a clause about any of them forfeiting all rights if he contests the will. Another clause allowing Leadburn to disqualify any of them for 'discreditable conduct.' No definition given."

"Fairly had 'em by the short hairs," the sergeant admitted. "No wonder they disliked him."

"Must have been feathering his nest to the tune of over three thousand pounds a year," the inspector pointed out. "He pocketed the surplus each year. Nice little nest-egg for his old age. And this arrangement had still five or six years to run. The youngster's only fifteen."

He thought for a moment or two, then put a question.

"That girl of Jack Sparkford's; has she any money?"

"Not a stiver," the sergeant declared emphatically. "She's as poor as a church mouse. I fished that out quite definitely."

"So Jack Sparkford would have to wait six years before he could get into double harness. H'm! And now he can get spliced to-morrow, if he wants to. Possible motive there," the inspector concluded thoughtfully "He was in the house, handy, that night. And there was that quarrel the maid overheard, on this very point of an extra allowance. H'm! Put a query against his name, I think. That finishes the insiders."

"The outsiders are Timothy Sparkford, Corfe, and some person or persons unknown," Sergeant Longridge suggested, entering into the spirit of systematic inquiry.

"What about Timothy Sparkford then? Why doesn't he live with the rest of them?"

"He won two or three hundred last year on a share in a Sweep ticket," Longridge explained. "As soon as he got that money he cleared out and went to live by himself in a cottage in Moor End Road, across the railway from Leadburn's place, with the sidings in between. I don't blame him; I'd have cleared out myself if I'd been in his shoes. It was no life for a man of thirty, under the thumb of old Leadburn."

He paused momentarily and then added: "A bad lad, Timothy. Wine and women; a short life and a gay one: that's his motto. He's got consumption, poor beggar! and Leadburn wouldn't pay for sanatorium treatment in the early stages; said he didn't believe in it. Just a fad, by his way of it. So he saved money. Timothy didn't love his uncle much; I could see that with half an eye."

The inspector looked up sharply.

"'Wine and women,' eh? That fits in with the talk about discreditable doings that the maid heard. And 'discreditable conduct' are the words in the will. Old Leadburn must have been threatening to disqualify Timothy. That meant one less to share in the capital in the final divvy-up. Something in that, perhaps."

"He's got an alibi though," the sergeant pointed out. "He called in Dr. Ackworth, just as he told me. I've checked that. Still," he added ruminatively, "alibis aren't always sound. And he bought liver that morning. He fetches his own stuff from the butcher, living alone as he does.

"The butcher remembered him buying liver that day; and the dog was poisoned with liver. But Corfe bought liver, too, that day—so I fished out from his butcher. I wonder the dog would touch stuff with that smell on it; but it seems they fed it only in the morning—it being a watchdog—so probably it was ready to bolt anything by the time it came to midnight."

"Most likely," the inspector concurred. "But why should Timothy want to poison his own family dog which wouldn't interfere with him?"

"Corfe's more likely for that," the sergeant admitted. "And I've fished out that Corfe bought some cyanide that evening. I've seen the entry he signed in the poison-book of the druggist who sold it to him. And Corfe's coat was torn a bit when I saw him on the morning of the murder. The dog hated him, so they say. But he swore then that he'd gone to bed at ten and never waked up till the morning."

"He lied then," said the inspector. "Somebody saw a light in the lodge-room about midnight. Got a note of it here. You told him to call in just now, didn't you? See if he's turned up and we'll put him through it."

In a minute or two the sergeant returned with Corfe. The gardener had the coarse looks and powerful physique of a fine animal, but not altogether a good-natured one. Something in the eyes suggested a dangerous temper which might break out suddenly and furiously, though at this moment he seemed sullen and uneasy rather than angry.

"We want more information than you gave the sergeant, Corfe," the inspector began abruptly. "First of all, how did you get that tear in your jacket?"

"Caught it on a nail and tore it," Corfe declared sullenly.

"A nail doesn't make that shape of tear in cloth," retorted the inspector. "What had you for supper on the night of the murder?"

Corfe pondered for some seconds before answering, as though he were weighing alternatives.

"Sausages," he said at last.

"So you had the raw liver in the house? You bought liver at the butcher's that day."

Corfe had the wit to see the trend of this. Caesar had been poisoned with raw liver. He corrected himself clumsily.

"My mistake. It was liver I had for supper, now I think of it. I ate the lot."

"What did you buy cyanide for?" the inspector continued.

This time the answer came promptly enough. "To kill rats with. They're in my chicken run."

"Dangerous stuff to have lying about," commented Dronfield. "Could the dog have got at it?"

"No, it was inside my fence. I put it at the rats' holes, not in the run, of course. I'm not a fool."

"You are, in some ways," Dronfield said acidly. "Look here, my man, I advise you to tell the truth. It'll do you less harm than the lies you've given us. That's a plain warning. We know a bit more than you think. You'd better come across."

Corfe shifted uneasily from one foot to the other as he digested this advice. He was so long in making up his mind that the inspector grew suspicious.

"You needn't start making up a yarn," he said sharply. "If you're going to tell the truth, it won't need any thinking over. Come along now."

Corfe pondered for a few moments longer. Then he seemed to have his story ready.

"This was the way of it," he began hesitatingly, like a man not too sure of his ground. "After the row I had with that old blackguard Leadburn, I had to have a talk with Jenny, naturally. But when I went for to see her, she was busy with their dinner and we couldn't get more than a word.

"I do a bit of jobbing work in the evenings to make some extra money, and I had to go to Mr. Rigg's in Broomhill Drive, and I didn't expect to be back till after ten. Old Leadburn wouldn't have a maid outside his door after ten o'clock. So I slipped a word to Jenny to be at the window in the hall upstairs at midnight and I'd come up, so as to have a talk with her about things. I thought I'd be able to get up on a bit of wall there, and keep clear of the damned dog. I was about beyond caring about dogs, then, in the state of mind I was in.

"So when I got back from Broomhill Drive I hung about a bit, waiting till it was time, and then I went up for to see her. But the dog beat me. I took a stick with me; but stick or no stick, it near had me down. It tore my coat for me and I had to give up. If I'd managed to get past it, it'd have raised Cain anyhow, barking, and had the house all awake. Lucky it fought quiet, so there was no row, barring the growling, and that roused nobody. But I had to turn back and go home to the lodge. And that's the plain truth, believe it or not."

"Did you see a light in the room with the French window?" the inspector asked, without commenting on Code's statement.

"The curtains were drawn, but the light was burning."

"When was that?"

"I went up just before twelve o'clock, as I told you," Corfe declared. "Just as I was going to bed, after I got home, I heard the express pass. That's twelve twenty-five."

"Very well," said Dronfield. "That'll do for the present."

He dismissed Corfe, and in a few minutes the sergeant ushered Jenny Hart into the room.

"You're quite sure about those things you told the sergeant?" was the inspector's opening. "About the quarrel amongst the three men on the night of the murder?"

"I heard them at it."

"You heard something about an allowance? And Mr. Leadburn said something about 'discreditable doings'? Sure of that?"

"That's what I heard."

"The murder took place at midnight. Where were you then?"

"Sitting in the upstairs hall, waiting for Simmy. He said he'd come if he could get past the dog."

"Hear anything while you were there?"

"Yes. Caesar gave a funny howl—sort of like a squeaking balloon it was, only louder, of course. Long-drawn-out sort of noise.

And I was afraid, a bit, about Simmy. I waited for a while in case he turned up. When he didn't, and I heard the express go past, I took it that Simmy wouldn't be up that night, so I went to bed."

"Any bedrooms on that floor, opening off the hall?"

"Master Sydney's room opens off it on one side and Mr. Jack's on the other."

"Moonlight night, wasn't it? Notice if the bedroom doors were shut?"

"Master Sydney's was. Mr. Jack's was open. I thought he hadn't gone to bed. When I passed his door I saw his bed hadn't been slept in"

"So Jack Sparkford was out of his room until past midnight," the inspector mused, after Jenny had been dismissed. "That's a bit of fresh news. And that little piece seemed to be speaking the truth then, which is more than Corfe did. Let's think it over again. Start with what we've got."

The sergeant checked over the facts on his fingers.

"There's the dog poisoned with cyanide; and the raw liver; and the knife; and the unfinished writing; and the burned newspaper; and the left-handed cut in his throat; and the tear in Corfe's coat; and Jack Sparkford up until the small hours; and the quarrels, of course," he ended rather vaguely.

"That seems the lot," Dronfield agreed. "Well, it wasn't either of the maids. The cook has no motive; and Jenny Hart could get Corfe to do the job sooner than do it herself—unless they were both in it together."

"The boy had cyanide in his butterfly bottle, and he knew about the knife being on the table," the sergeant suggested.

"He'd no motive and no raw liver," the inspector objected impatiently.

"There's Corfe. He's been telling a pack of lies and he did get his coat torn by the dog that night. And he'd both liver and cyanide in hand. And *he* had a motive, right enough."

"Leave him aside for a moment. See how the rest of them stand."

"There's Timothy Sparkford," Longridge suggested rather doubtfully. "I'd have put my shirt on him as the one that did it.

He's got the temper for it, or I'm a Dutchman. If old Leadburn meant to use that clause about discreditable conduct against him, Timothy was going to be cut out altogether after waiting so long for his share. There's motive enough. And the bit of writing might be something about the cutting-out that Timothy just stopped in time. And the left-handed cut fits him, with his left-handedness. And the burnt newspaper may have been a bit that he used to wrap round the handle of the knife to keep his finger-prints off it; and he had to burn the paper afterwards on the same account."

"Anybody might have done the same," the inspector objected. "That doesn't fit Timothy specially."

"Then he bought some liver that day," Longridge continued, disregarding the interruption. "In fact, as I say, I'd have put my shirt on him. Only, it won't wash," he added regretfully.

"Are you so sure as all that?" the inspector demanded.

"It won't work," Longridge insisted. "You know the lie of the land about the house. To get from it to Timothy's cottage you've got to cross the railway. There's two bridges: one a quarter of a mile north of the house, and t'other one half a mile to the south. That makes it either half a mile or a mile by road from door to door.

"The dog was poisoned at 12.20 a.m. Besides, what would he poison the dog for, seeing he could walk past it without bother, seeing it was friendly? Well, it was poisoned at twelve-twenty. Then there was some time spent in doing the murder itself—ten minutes, at least, and likely longer. That makes it 12.30 a.m. before Timothy could start off home again.

"But it was just about half-past twelve when he rang up Dr. Ackworth from his cottage; and the doctor got there very soon afterwards to find him ghastly sick. I've seen Dr. Ackworth and them's the facts. Besides, he couldn't have got across either bridge without being seen, as it happens. The road was up at the north bridge, and there was a watchman on all night who swears that between eleven and two o'clock nobody passed him, barring a tall, slim young fellow in plus-fours, bare-headed, and wearing a white scarf.

"That won't fit Timothy. He's short, with a figure like a gorilla. On t'other bridge there was a motorist in trouble, kept there for over an hour between eleven-thirty and twelve-thirty o'clock, fixing something under a street lamp. He and his passenger are certain nobody passed, bar one man: the same cove in the white scarf. He stopped to ask if he could lend a hand. That was about a quarter-past twelve, they say. Timothy's easy remembered, and he wasn't seen at either bridge."

"Well, what was to hinder him going straight across the railway line, making a bee-line from house to cottage?"

"It won't work," Longridge protested in an aggrieved tone. "Timothy suffers from something they call comical cornea. He's as blind as a bat, even on a bright moonlight night like that one.

"Now look how the land lies. There's a rock cutting from the one bridge to the station. He couldn't have climbed down that rock face and up t'other side, not with his sight. It's impossible; let alone there isn't the time. He couldn't get into the station. It's locked up at that time of night, and it's a solid block of building on the lip of the cutting. You can't get at the stair down to the platform without breaking in the door of the ticket-office. He didn't go that way.

"Beyond the station, just opposite the house, there's a barbed-wire fence on level ground, and a mass of sidings filled with odd trucks standing, and beyond the sidings there's another stiff fence on the Moss Cottage side of the line.

"I might be able to get across by that route in ten minutes myself, though it would be quick work: but a man with Timothy's comical cornea simply couldn't come near that time. He'd be absolutely lost among the trucks, let alone that he couldn't shin over the fences like a normal man who could see what he was doing.

"I've been over the ground myself. He simply couldn't have managed it. And farther north, beyond the sidings, there's another rock-cutting, sheer in the sides, that would take a man with all his sight to get down. No half-blind man did that job, you can take it from me."

"H'm!" said the inspector, convinced by this evidence, "then that leaves Jack Sparkford. But he doesn't fit in anyhow. It's his own dog, so he needn't be afraid of it interfering with him. Besides, he's in the house already and doesn't need to pass the dog."

"He had a motive," the sergeant insisted. "If old Leadburn was out of the way he could marry the girl. Maybe he was putting the screw on the old man and went a bit too far—killed him without actually planning to do it.

"Making him write out, 'I hereby agree to give Jack Sparkford an allowance fit to marry on,' or something of that sort. That would fit the facts. And he *was* up and about, late that night, by the girl's evidence. He kept his thumb on *that* bit when he was giving his own account of things."

"We'll go over and see him now," the inspector decided, after a glance at his watch. "Don't like his keeping back information this way. Not good enough."

They found Jack Sparkford at home, and after the sergeant had introduced his superior, the inspector opened the matter with his usual bluntness.

"You told Sergeant Longridge that you went up to bed at 11 p.m. on the night of Mr. Leadburn's death, sir. Was that a slip, by any chance? Would you like to correct it?"

Jack seemed taken aback by this suggestion.

"What do you mean?" he demanded, rather uncertainly.

"You went up to bed about eleven," the inspector conceded. "But did you get into bed?"

Jack seemed to be all on the alert.

"What I said was perfectly accurate," he insisted. "I went upstairs at eleven o'clock."

"Yes, yes," said the inspector testily. "But what did you do after that? You didn't go to bed immediately. What were you doing?"

Jack paused for a moment or two, then he seemed to come to a decision.

"You seem to know something. I've nothing to conceal, so I'll be quite frank with you. I'd had a row with my uncle, a private matter, nothing to do with his death. He wouldn't come round to my view. It was an important matter to me. I saw my brother part of the way home when he left at ten o'clock. Then I came back again and tried to persuade my uncle again.

"It was no use; he was quite set in his view. So I went upstairs at eleven o'clock. I was worried, very worried, when I got up to my room. I knew it was no good going to bed, I'd never have slept. I sat about in my room for a while, about an hour, I should guess, thinking. Then I came downstairs and went out of the house. I wanted to walk off my troubles."

"What time was that?" the inspector demanded.

"I didn't look at my watch."

"How were you dressed?"

"Plus-fours—the same as I have on now. And I put on a white scarf because I'd had a touch of sore throat that day."

Jack unconsciously clinched the matter with his next words.

"I remember one thing that may help you," he added with something like a sneer. "On the bridge beyond the station I came across a fellow under a street lamp with a twelve-six Austin car. He'd been caught with a puncture with his spare wheel out of action. He'd patched up the puncture with a portable vulcaniser, put his wheel on again, and just as I got up he found the nail had gone through both sides of his inner tube, and he had missed the second hole when he had it down. He was just starting to take it off the wheel a second time when I passed him and offered to give him a hand. If you can get hold of him, he'll identify me, I expect. It must have been round about midnight when I had my talk with him."

"Which way did you come home again?" the inspector asked.

"By the other bridge. I don't know when I got in, certainly long after midnight."

"Did you see any light in the room with the French window as you came back?"

"No, not that I remember. I didn't look particularly. I'd left my uncle making up his accounts."

"You didn't see the dog Caesar?"

"I saw him on my way out. He went part of the way down to the gate with me, and then I ordered him home. I didn't see him on my way back, but I thought nothing of that. By the way, I remember that the fellow with the Austin car glanced at his watch once while I was with him. He may remember what time it was, then, if you're really interested."

"Can you suggest any explanation of the uncompleted document your uncle was writing that night?" inquired Dronfield.

Jack shook his head.

"I haven't the faintest idea."

The inspector, quite satisfied, withdrew along with his subordinate. As they walked together down the drive, the sergeant broke silence.

"That's the lot of them cleared, if you take the evidence as sound."

"Corfe's story's unsupported except by that girl of his; and her evidence doesn't tell us what he really did," objected Dronfield. "It's on the cards that the two of them were in it, and just cooked up a yarn between them."

"Might be that," the sergeant agreed, thoughtfully.

"I've got to hurry off now," Dronfield said, with a glance at his watch. "I'll just save that appointment and no more."

"Merely a slight attack of conjunctivitis," the specialist assured Inspector Dronfield. "Nothing to worry about, though it's been a nuisance to you, I expect. I'll write you a prescription for some eye-drops."

He went over to his writing-table and jotted down something on a sheet of paper.

Meanwhile the inspector, left to himself, let his eyes wander over the various appliances of the oculist's armoury: the box of lenses, the ophthalmoscope, the perimeter, the astigmometer, and the case of test types with its concealed lamp. Finally his glance fell upon something which puzzled him by its very simplicity: a little disc, concentrically ringed like a target, with a hole where the bull's-eye should have been, and furnished with a handle like a lorgnette.

The inspector was the last patient on that day's list; and the oculist was a man who liked to relax after his work was done. As he came back with his completed prescription, he noticed Dronfield's interest in the little instrument.

"Looking at the Placido's disc?" he inquired with a smile. "I'll give you three guesses and see if you can spot what we use it for."

"I was just wondering, but I didn't get the length of guessing," the inspector confessed. "It's beyond me. What *is* its use?"

"It's for diagnosing conical cornea," the specialist explained, picking it up as he spoke.

The inspector became alert. Conical cornea! Of course, that was what the sergeant had misheard and turned into "comical cornea." The trouble with Timothy Sparkford's eyesight. The inspector decided that there would be no harm in hearing more about that subject.

"Could you explain it, sir?" he asked, with obvious interest. "What *is* conical cornea?"

"It's a malformation of the eye. You know what the cornea is? The transparent covering at the front of the eyeball. In the normal eye, it's roughly spherical, and it acts as part of the mechanism of sight. If it's misshapen—conical in form instead of spherical—it distorts vision. Unfortunately you can't correct the distortion with spectacles."

"Not even with special spectacles? I see. And where does the little target-thing come in, sir?"

"It shows up the defect at once. I bring it up to the patient's eye, target side towards him. The rings are reflected in his cornea, as if it were a convex mirror; and by looking through his hole where the bull's-eye should be, I can examine the reflection. If he's got conical cornea, the reflection's distorted in a peculiar way that's recognisable at a glance. Very neat, isn't it?"

"Very neat indeed," the inspector acquiesced. "And you can't cure it with spectacles?"

"No, spectacles are of no use. In fact, until lately, it's been hopeless."

"Until lately?" demanded the inspector sharply. "You mean it can be cured now?"

The oculist leaned over and opened a box which lay on his table.

"Curiously enough, I had a patient here to-day to be tested for these things. They're what are called 'contact glasses.'"

He showed Dronfield a series of tiny objects, almost hemispherical in form and made from glass so thin that they seemed the most fragile things Dronfield had seen.

"The idea is this," the oculist went on. "The essence of the trouble is that the outer surface of the patient's cornea is conical instead of spherical. Fill one of these little glass cups with saline solution and slip it under the eyelids, above and below. It sticks to the eyeball by surface tension, and the salt water fills up the gap between eye and glass.

"The net result is that you've now got a glass outer surface to your eyeball, and that new surface is shaped just like a normal cornea. You've merged the defective conical surface into a system with the same refractive index . . . . but perhaps that's getting a bit too technical for you."

"And a man wearing one of these gadgets sees as well as a normal person?"

"So far as his conical cornea goes, yes. And what's more, no one would ever imagine that the patient was wearing anything. It's not like spectacles. These contact glasses, once they're in place, are almost unnoticeable, unless you've been told to look for them."

"I suppose they take some getting in, though?"

The oculist shook his head.

"Not a bit of it! The patient can do that for himself; insert them in the morning and take them out again when he goes to bed, if he wishes. They hardly cause the wearer any discomfort."

"I see, sir. That's wonderful. Very interesting indeed, sir. I'd no notion things of that sort were possible. And now I mustn't detain you any longer. This the prescription? Thank you, sir."

And with as much haste as courtesy allowed, the inspector bowed himself out of the consulting-room. His first call was at a druggist's, where he asked one or two questions. Then he returned to his headquarters and summoned Sergeant Longridge.

"We've got the Leadburn murderer at last," he said, as his subordinate entered the room.

"Can you prove it?" the sergeant asked sceptically. "Who is it?"

"I think so, if we've any luck in a search of the premises after we arrest him. It's Timothy Sparkford."

"But the thing's flat impossible," Longridge protested. "I've checked all the evidence to the last dot, in his case, because at the start I was dead sure he was the man we wanted. Even if you leave out his sight—and that makes the affair impossible in itself—the times won't fit. Nohow. See here. The dog was poisoned at 12.20 a.m.

"After that, he had to do the murder, get away, climb a fence, cross the siding, climb the other fence, and get to his cottage, all by 12.30 a.m., for he telephoned to Dr. Ackworth about that time. I've timed myself over that course, going as hard as I could, and it took me nine minutes from door to door. Where's the time required for murdering old Leadburn?

"Besides, Timothy had eaten some bad tinned stuff that evening and was deathly sick. We've Ackworth's evidence for that. A man as sick as all that simply couldn't commit a murder neatly and then do all the gymnastics required to get home in double-quick time, let alone he's as blind as a bat."

"You can wash out the 'blind as a bat' part," Dronfield declared. "He's only blind when he chooses to be."

And he explained the matter of the contact glasses to his subordinate, who opened his eyes at the information.

"Amazing what they can do," he admitted.

"Still, you can't get over the rest of the facts."

"Let's take it step by step," suggested Dronfield, who prided himself on having a systematic mind.

"Here's how I figure it out. Timothy's been thinking of this for a good while. He goes up to London and gets himself fitted with contact glasses. That leaves no clue among the eye-specialists hereabouts. Of course he says nothing about the glasses to anyone. That equips him with fair normal sight, unknown to anyone in these parts. Now come to the day of the murder. He buys some liver. We know that. He's got cyanide to hand. . . ."

"How?" demanded Longridge.

"From his electro-plating hobby. Cyanide's used in silver-plating. That fits him out for the dog-poisoning. He goes up to the house that evening to give his brother Jack a helping hand, and to make sure that all's favourable to his plans up to the last possible minute. Then he clears out at 10 p.m."

"That seems straight enough," Longridge admitted. "It's the next stage that's sticky."

"Well, he clears out at ten o'clock, and Jack has to see the poor blind bat half-way home. That's bound to come out in evidence, and it impresses his bad sight on simple fellows like you and me.

"Later on, he puts on his contact glasses and comes back. I don't know when, exactly. Just before midnight, probably, so as to catch old Leadburn before he finishes up his accounts. He passes the dog without its barking. One of the family. He tries to force old Leadburn into signing some document, probably something that would put an end to the avuncular tyranny.

"Method of persuasion: a cord round the neck. No chance of Leadburn yelling for help in these circs. Unfortunately, the cord gets drawn a bit too tight. Leadburn chokes. Timothy tries to cover up the cord marks by cutting the old man's throat. He holds the knife in a bit of newspaper and burns the paper afterwards. No finger-prints, in that way. But in the flurry he forgets all about left-handedness and leaves that clue for us. Then he goes off, calls up the dog, poisons it with cyanided liver. . . ."

"Why?" demanded Longridge. "I don't see the point, there."

"I see two," retorted the inspector. "First, the dog-poisoning makes it look like an outsider's job. Second, beasts poisoned with cyanide give a loud cry; and Timothy takes care to lead the beast under his brother's window to kill it. The youngster's got toothache and isn't sleeping sound. The howl wakes him. That dates the poisoning at twelve-twenty all right, for Timothy would bear the express in mind. And we dropped into his trap and assumed the dog had been poisoned by an intruder on his way in to the house. Makes all the difference in the supposed timing of the affair. Gained Timothy all the time he took in the actual murder, see?"

"I see," said Longridge. "Go on."

"With his contact glasses on, he'd get to the cottage as quick as you could. Nine minutes, you said. He'd be home by 12.30 a.m. Then at once he rings up Ackworth and says he's been sick. He's no more sick than you are. But as soon as he's done phoning, he swallows an emetic."

"Ipecacuanha?" queried the sergeant, conquering the polysyllable by careful enunciation.

"No, it's too slow in acting. Copper sulphate would do the trick. It acts immediately. And he'd have it for his electro-plating stunt. By the time Ackworth arrives, Timothy's sick enough; and he tells that lie about some tinned stuff having given him the gripes. Naturally Ackworth suspects nothing. So there's a sound alibi established. See?"

"It sounds neat," Longridge admitted. "And it clears Corfe of the dog-poisoning, which was the sticker from my point of view. You're going to get a warrant?"

"Yes. You'll execute it. Take a man or two with you and comb his cottage thoroughly after he's in custody. What we want is cyanide, copper or zinc sulphate—they're both emetics—his contact glasses, any bills or papers you can find that bear on his purchase of the glasses. If you get these things, we have him by the short hairs, I think."

"You'll have to suggest some motive that'll pass with a jury," objected the sergeant.

"No difficulty about that," declared the inspector. "We know our Timothy. Great lad for wine and women. What about the discreditable conduct clause in the will? Besides, his lungs are gone, aren't they? He's thirty, now. I don't know how far gone he is, but most likely he thought he wouldn't be in any condition to enjoy his fortune if he had to wait six years more for it. 'A short life and a gay one,' was his idea. But you can't be gay on £100 a year—not in that line of gaiety, anyhow. So it was a race between his consumption and the date of his inheritance. All he wanted was to shift that date a bit forward, so that he could enjoy his money while he was still fit for it. That's how I see it, anyhow."

"Lucky you had to see that eye-doctor," was Sergeant Longridge's reflection, which he kept to himself.

Coachwhip Publications
CoachwhipBooks.com

ISBN 978-1-61646-317-1

Coachwhip Publications
CoachwhipBooks.com

ISBN 978-1-61646-320-1

Coachwhip Publications
Also Available

ISBN 978-1-61646-321-X

Coachwhip Publications
CoachwhipBooks.com

THE BOAT HOUSE RIDDLE
J. J. CONNINGTON

ISBN 978-1-61646-306-6

Coachwhip Publications
Also Available

ISBN 978-1-61646-310-4

Printed by Libri Plureos GmbH in Hamburg, Germany